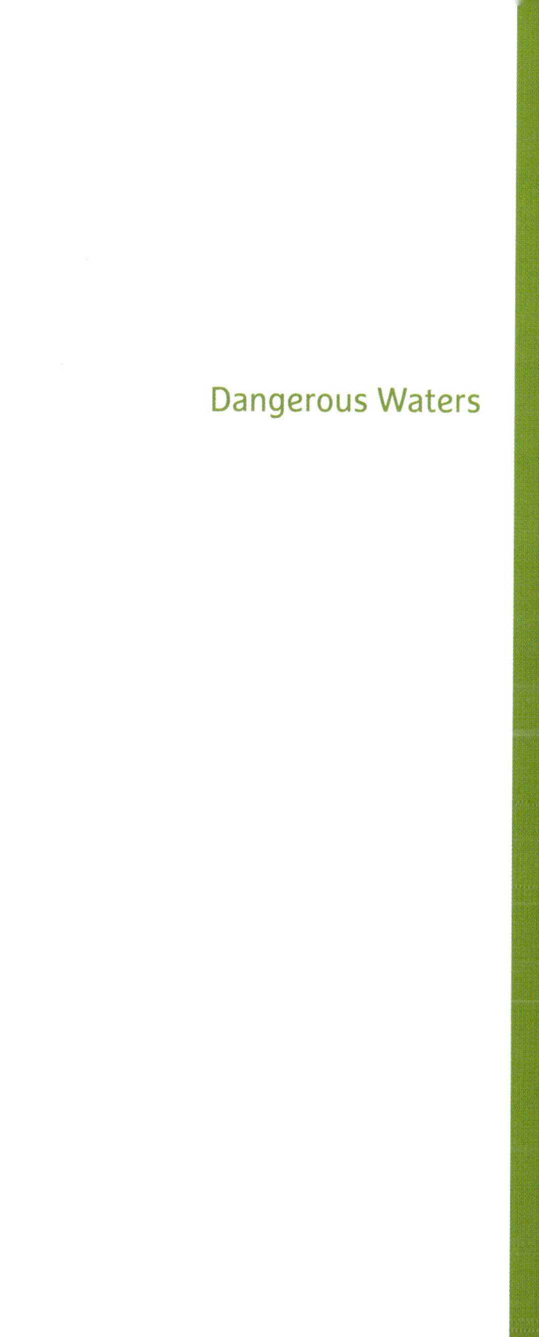

Dangerous Waters

DANGEROUS WATERS

A Photo Essay on the Tennessee Valley Authority

MICAH CASH

The University of Tennessee Press / Knoxville

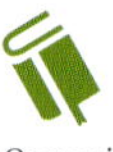

Library of Congress Cataloging-in-Publication Data

Names: Cash, Micah author, photographer.

Title: Dangerous waters: a photo essay on the Tennessee Valley
Authority / Micah Cash.

Description: Knoxville: The University of Tennessee Press, [2017] |
Includes bibliographical references.

Identifiers: LCCN 2016053772 | ISBN 9781621903574 (hardcover)

Subjects: LCSH: Tennessee Valley Authority—History. | Tennessee
River Valley—Pictorial works. | Flood dams and reservoirs—
Tennessee River Valley—Pictorial works.

Classification: LCC F217.T3 C37 2017 | DDC 976.8—dc23

LC record available at https://lccn.loc.gov/2016053772

TO ANNE

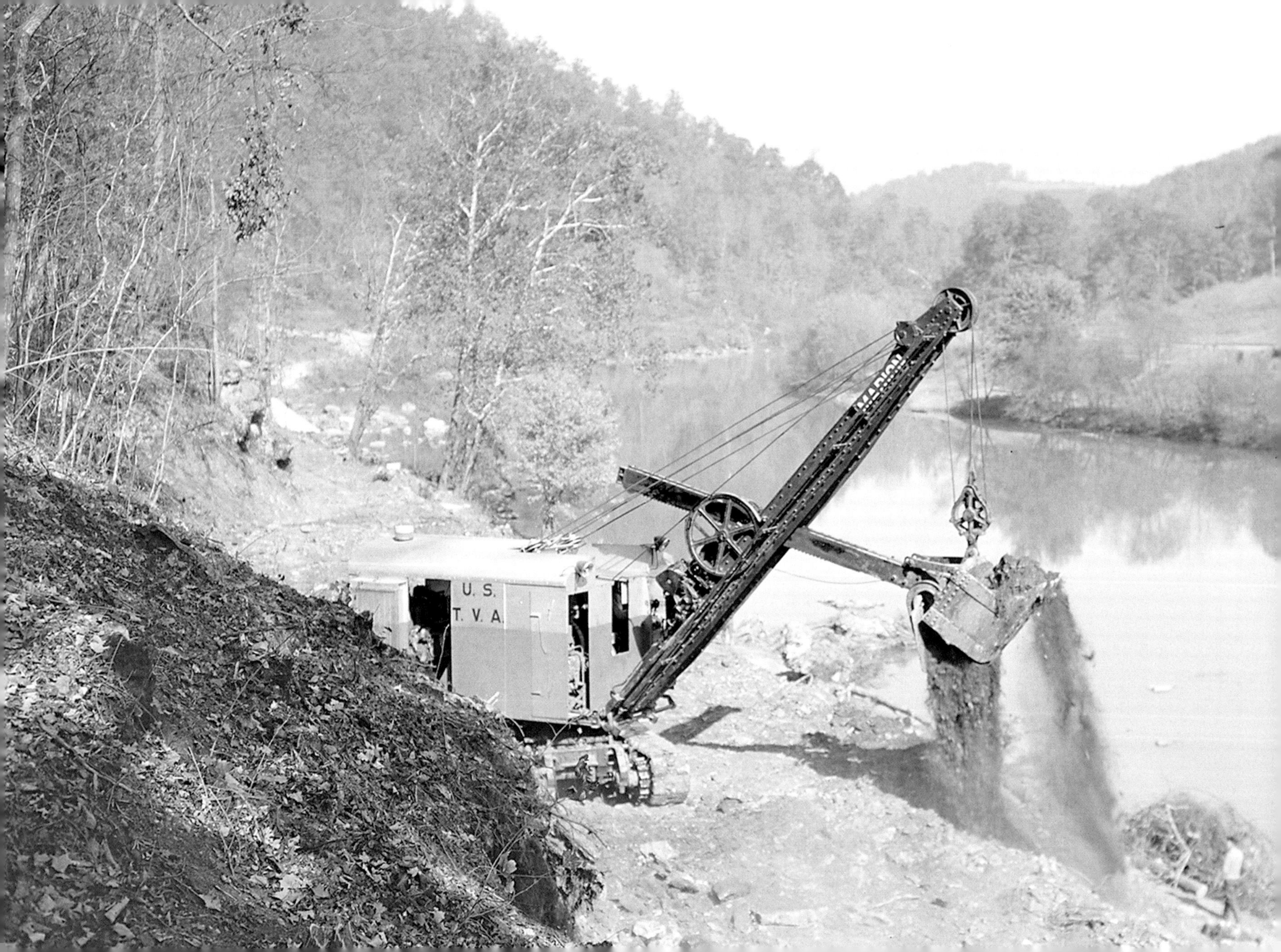

U. S.
T. V. A.
MARION

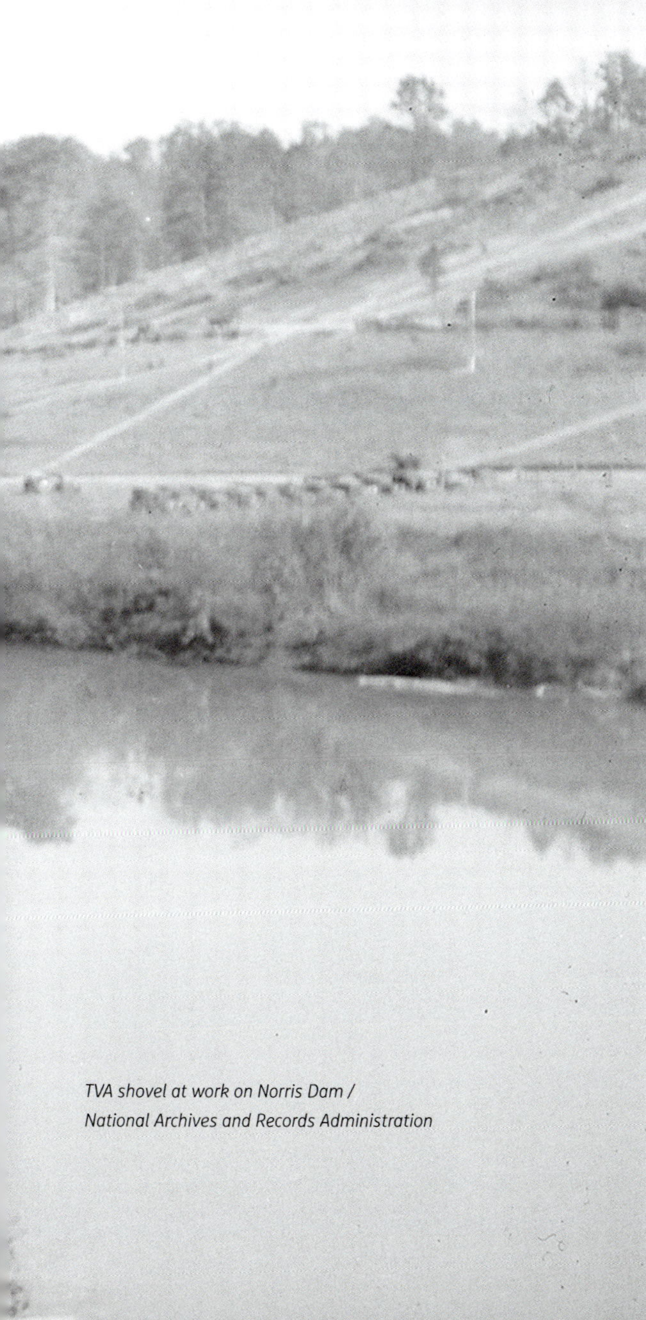

TVA shovel at work on Norris Dam /
National Archives and Records Administration

Contents

Acknowledgments

What began as a simple conversation with friends and mentors about history, hydroelectric power, and landscape slowly grew into a multi-year project that was aided by dozens of people.

The original fieldwork for *Dangerous Waters* was supported by the Zachs Endowment Fund from the University of Connecticut. I want to thank everyone at the University of Connecticut School of Fine Arts for their support, especially Janet Pritchard, Alexis Boylan, Kathryn Myers, Deborah Dancy, and Judith Thorpe. I would also like to thank Daniel Buttrey for his valuable advice, keen eye, and technical exper-tise. This group of talented artists and exceptional educators provided necessary feedback when this project was in its infancy and helped sharpen my photographic inquiry as I completed subsequent fieldwork.

My archival research was aided by Maureen Hill and Joel Walker at the National Archives at Atlanta. They followed the project from an academic inquiry to a complete body of work. Their invitation to participate in the 2014 symposium, "Valley of the Dams: The Impact and Legacy of the Tennessee Valley Authority" provided an opportunity to contextualize this work alongside that of current TVA scholars. In addition, I want to thank Pat Bernard Ezzell and Nancy Proctor from TVA. I met both during the summer of 2013 at the TVA library in Knoxville, Tennessee. They were consistently helpful in providing historical imagery and documents during the project's early stages.

Many of the images from this book have been featured in publications over the years, and I am thankful for those editors and publishers who believed in the project and wanted to share it with their readership. A sincere thank you to Chuck Reece, David Whitling, and Kyle Tibbs Jones of *The Bitter Southerner*; Nancy Levinson, Josh Wallaert, and Tim Culvahouse of *Places Journal*; Ashley Kauschinger of *Light Leaked*; Lauren Walton of *Ninth Letter*, Ian Sarjeant of *Another Place*, Ryan Sparks of *Southern Glossary*; and Carson Sanders and Taylor Curry of *Aint-Bad*. Each of these individuals are supporters of photography and believers in storytelling, and I am honored to have been featured in their periodicals.

I have been fortunate to exhibit much of this work over the course of this project and would like to thank Steven Pearson at McDaniel College in Westminster, Maryland; Maureen Hill and Joel Walker at the National Archives at Atlanta; Taylor Petree; Robin Bernat at {Poem 88} in Atlanta, Georgia; Steven Skopik at Ithaca College, New York; and Glen and Maria Nocik at C3 Labs in Charlotte, North Carolina.

My sincere gratitude to people who made this book happen: the amazing and trusty Christine Fadden who provided final edits to the manuscript and to Jean Nihoul who read over the first draft; to Thomas Wells at the University of Tennessee Press for his vision for this book and his guidance and patience throughout the process, and to Kelly Gray for the thoughtful design.

I am grateful to the friends and colleagues who have offered guidance, support, and friendship over the years: Leslie Booren, Roger May, Molly Lamb, Ethan Fogus, Jeff and Dyana Barninger, Brianna Bedigian, Rob Morgan, Beth Lyons, Emily Rafferty, Todd Stewart, Katie Lloyd, Amy Bagwell, Amy Herman, Graham Carew, Brandon Mathias, Jean Nihoul, Lindsay Simon, Matt Bowman, Julia DePinto, Shane Morrissey, Allison Hale, Reagen Holt

O'Reigaekn, Jared Holt, and Kathleen Deep. I want to also thank my family, both immediate and extended, for their consistent enthusiasm and overwhelming support.

My deepest thanks go to my wife, Anne Henry Cash, whose unwavering support was a guiding force behind these photographs. She was with me when I made the first images and witnessed the birth of this project, never questioning the journey it took us on.

DANGEROUS WATERS

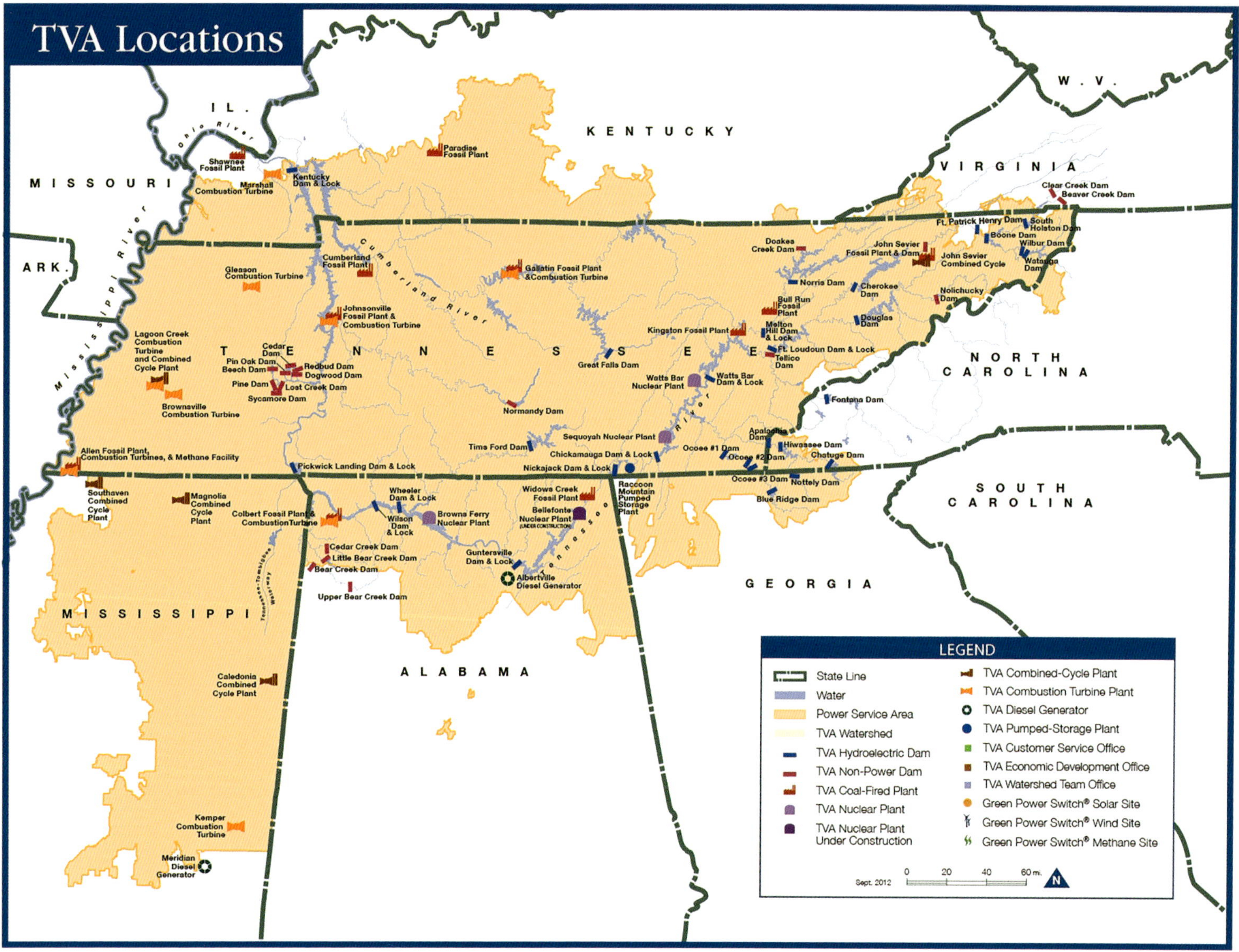

TVA Locations

MISSOURI
ARK.
IL.
KENTUCKY
VIRGINIA
W. V.
TENNESSEE
NORTH CAROLINA
SOUTH CAROLINA
GEORGIA
ALABAMA
MISSISSIPPI

Ohio River
Mississippi River
Cumberland River
Tennessee River
Tennessee-Tombigbee Waterway

Shawnee Fossil Plant
Kentucky Dam & Lock
Marshall Combustion Turbine
Paradise Fossil Plant
Clear Creek Dam
Beaver Creek Dam
Ft. Patrick Henry Dam
South Holston Dam
Boone Dam
Wilbur Dam
Watauga Dam
Doakes Creek Dam
John Sevier Fossil Plant & Dam
John Sevier Combined Cycle
Gleason Combustion Turbine
Cumberland Fossil Plant
Gallatin Fossil Plant & Combustion Turbine
Norris Dam
Cherokee Dam
Nolichucky Dam
Bull Run Fossil Plant
Douglas Dam
Johnsonville Fossil Plant & Combustion Turbine
Kingston Fossil Plant
Melton Hill Dam & Lock
Ft. Loudoun Dam & Lock
Tellico Dam
Lagoon Creek Combustion Turbine and Combined Cycle Plant
Cedar Dam
Pin Oak Dam
Beech Dam
Redbud Dam
Dogwood Dam
Pine Dam
Lost Creek Dam
Sycamore Dam
Great Falls Dam
Watts Bar Nuclear Plant
Watts Bar Dam & Lock
Fontana Dam
Brownsville Combustion Turbine
Normandy Dam
Apalachia Dam
Hiwassee Dam
Chatuge Dam
Allen Fossil Plant, Combustion Turbines, & Methane Facility
Tims Ford Dam
Sequoyah Nuclear Plant
Ocoee #1 Dam
Ocoee #2 Dam
Pickwick Landing Dam & Lock
Chickamauga Dam & Lock
Nickajack Dam & Lock
Ocoee #3 Dam
Nottely Dam
Southaven Combined Cycle Plant
Magnolia Combined Cycle Plant
Wheeler Dam & Lock
Widows Creek Fossil Plant
Raccoon Mountain Pumped Storage Plant
Blue Ridge Dam
Colbert Fossil Plant & Combustion Turbine
Wilson Dam & Lock
Browns Ferry Nuclear Plant
Bellefonte Nuclear Plant (UNDER CONSTRUCTION)
Cedar Creek Dam
Little Bear Creek Dam
Bear Creek Dam
Guntersville Dam & Lock
Caledonia Combined Cycle Plant
Upper Bear Creek Dam
Albertville Diesel Generator
Kemper Combustion Turbine
Meridian Diesel Generator

LEGEND
State Line
Water
Power Service Area
TVA Watershed
TVA Hydroelectric Dam
TVA Non-Power Dam
TVA Coal-Fired Plant
TVA Nuclear Plant
TVA Nuclear Plant Under Construction
TVA Combined-Cycle Plant
TVA Combustion Turbine Plant
TVA Diesel Generator
TVA Pumped-Storage Plant
TVA Customer Service Office
TVA Economic Development Office
TVA Watershed Team Office
Green Power Switch® Solar Site
Green Power Switch® Wind Site
Green Power Switch® Methane Site

Sept. 2012
0   20   40   60 m.
N

Dangerous Waters: Signs proclaim this warning at every Tennessee Valley Authority (TVA) hydroelectric dam and are a stark reminder that these locations are not as serene or controlled as they initially seem. The powerhouses create intense currents, and waters can churn unpredictably and rise at any moment. The billboards serve as a direct warning to boaters and anglers and confirm that the recreational opportunities enjoyed by all come with an inherent risk. In addition, they call to mind the actions taken in order to bring power to the Tennessee Valley region: the flooding of fertile land and the displacement of families. No other name could better describe this photographic project, literally and metaphorically. *Dangerous Waters* is lifted directly from the signs that mark these altered landscapes; it is also a metaphor that considers the history of these dams and acknowledges that TVA's decisions continue to be a divisive force within the region.

Landscape is an inherently conflicted visual and cultural space, and my interest as an artist lies within this charged arena. As geographer D. W. Meinig reminds us, a particular landscape is uniquely defined by the viewer.[1] Any single vista can be perceived in a variety of ways depending on our intention and lived experience. Land and water are viewed both as resources to protect and resources in which to mine and invest. As a society, we routinely make ecological, social, and economic compromises based on potential profit, preservation, or progress, and we cannot always foresee or plan for the consequences. The physical alteration and cultural reclassification of a particular landscape brings about change in a variety of deliberate and unintended ways.

When President Franklin Delano Roosevelt created the Tennessee Valley Authority in 1933, he did so because transformation of the Tennessee River watershed was necessary. Rural populations did not have electricity, annual floods ravaged the region, and large sections of the Tennessee River were difficult to navigate. The Valley was to be the birthplace of a social and political initiative that would necessitate one of the largest physical landscape transformations in American history. The burgeoning mindset was one of idealism and reassurance, employment and opportunity, and people before profit. TVA showcased the government as benevolent and promised affordable and abundant electricity.

In order to achieve such lofty goals, the topography of the Valley had to change. Progress required the building up of concrete and steel and the covering up of mountains and valleys with inland seas and waterways. Four generations later, the people that live within TVA's service area continue to be affected by Roosevelt's ambitions and the agency's storied dams.

Pervasive and powerful, TVA's twenty-nine hydroelectric dams maintain enormous influence on the culture and topography of the southern Appalachian region—and they were built for us, the people of the United States.[2] The dams actively sustain a thriving hydraulic network of industry, hydroelectricity, and recreation. Originally constructed as nationalistic symbols of progress and opportunity, they enabled affordable electricity and thus spurred industrial and economic development.

Over the decades, TVA has transformed into a fully diversified utility company, growing past its original hydroelectric program goals. Coal-fired power production took precedence in the 1950s and nuclear power followed a generation later. Shrouded in New Deal nostalgia, the dams are commonly referred to in a historical sense, yet they are still generating electricity, impounding water, and upholding

TVA's essential mission of social welfare. Both in social memory and cultural practice, the dams retain a reverent presence for much of the local populace.

This reverence stems from TVA's multi-faceted social mission, which in 1933 was, "to improve the navigability and to provide for the flood control of the Tennessee River; to provide for reforestation and the proper use of marginal lands in the Tennessee Valley; to provide for the agricultural and industrial development of said valley; to provide for the national defense by the creation of a corporation for the operation of Government properties at and near Muscle Shoals in the State of Alabama, and for other purposes."[3]

Clearly, the agency's reach extended well beyond the traditional role of a utility provider. Serving as a model for large-scale infrastructure investment and long-term regional development, the effects of its policies and initiatives have been both nationally disruptive and globally studied. While TVA electrified a portion of the rural southern United States, it also eradicated malaria within the region and played a significant role in powering a portion of the World War II weapons production effort. Today, TVA continues to manage a vast network of recreational land, serving millions of visitors annually, and serving as a catalyst for economic and community development. I have spoken with dozens of TVA employees who firmly believe in what they do and continue to stand by the original social mission of their employer. To understand TVA and its contemporary significance one must acknowledge an existing conflict and duality: The blue logo that is emblazoned on everything from polo shirts to street signs can signify to the local populace equal parts sadness and pride.

Nothing about the agency's history is simple: its very origins were audacious yet optimistic. For the betterment of a region, entire towns were sacrificed, generations of families were removed, 1.3 million acres were purchased,[4] and approximately twenty thousand graves were relocated.[5] Fertile river-bottom farmland was washed over in the name of progress and modernity. And yet, the recreational land TVA manages constitutes a civic contract to the descendants of the families that were relocated—and to the memory of inundated ancestral homesteads. TVA is a socialized government agency that enthusiastically supports a regional capitalistic economy; its architecture is meant to inspire by blending in while simultaneously commanding visual attention. Such juxtapositions exist throughout the agency's history and can still be experienced throughout its seven-state service area that includes the entire state of Tennessee, much of Mississippi, and sections of Alabama, Kentucky, Georgia, North Carolina, and Virginia.

The agency's service area is defined by geography rather than by politics or social deliberation. Its jurisdiction is that of the Tennessee River watershed, from the tributary dams in western North Carolina and eastern Tennessee to the convergence of the Tennessee and Ohio Rivers downstream of Kentucky Dam. It is a system of water and industry, meandering out of the lush Appalachian Mountains and punctuated by breathtaking lakes. Each reservoir is unique: some are secluded and surrounded by national forest and parkland, while others are encircled by residential and industrial development. Geological factors tint the waters a variety of colors, from emerald green to mossy brown, and during the recreational season, vistas are dotted with anglers, swimmers, boaters, and campers. And yet, reminders of the agency's tight grip over the landscape and industrial usage

are everywhere: Transmission lines radiate from the dams and across the valleys like floating highways of electricity, billboard-sized signs issue safety warnings and jurisdiction reminders to the public, and intake towers thrust up obtrusively from the water's still surface. The compromise is clear: these are landscapes of both industry and leisure.

Dangerous Waters investigates the contemporary landscape of TVA's hydroelectric program, capturing in images the tenuous balance between locations that are designed for both industry and leisure. Fully informed by TVA's history, my focus for this project was to present and question, in visual terms, how these highly engineered landscapes are culturally, economically, and socially defined and utilized four generations after their creation. Made over the course of three years, the photographs are as much a documentation of my journey through TVA's history as they are of the places themselves. In the process of capturing these places and the people in them, I grappled with the intertwining threads of nationalism, history, economics, and socioeconomics.

The book you hold in your hands is the culmination of a journey that began in June 2013, atop Norris

Dam, and ended in June 2016, at the Land Between the Lakes National Recreation Area. My initial plan for this project was to visit every flood control and hydroelectric power facility that is owned, operated, or built by TVA. My goal was to understand the agency's hydroelectric program as a complete experience and to understand how the architecture and tonality of these spaces has changed over the generations.

These photographs document the locations as they are today, not as they were during the Great Depression or at any other time in their history. Some TVA locations are colored more by their past than others, and each dam reservation is entirely unique: some haunting and quiet, others alive with people and machinery. I witnessed simple yet beautiful weddings, large family reunions and intimate first dates, slow drifting solo fishermen and raucous boat parties, all within the vicinity of the dams.

And yet despite the natural beauty and human activity I encountered at the more popular recreational areas, there is a lingering sense of melancholy at many TVA locations. Industrial embankments are uncomfortably close; the hum of electricity is ever-present; and signs, fences, and other barriers mark clear designations of where human bodies can and cannot go. Over the course of this project, I came to realize that TVA-controlled recreational locations most commonly service those on the lower end of the economic spectrum: the families that do not have access to private swimming pools or the means to store their boats at commercial marinas. The economic divide was sometimes visually harsh as lakes are often surrounded by beautiful homes and gated residential communities, which also means their extensive piers and lake-overlooks are inaccessible to those without membership status. This presented itself as a striking contrast to one of the main historical narratives of TVA: *Access.*

Eschewing the utilitarian appearance of traditional dam architecture, TVA designed their facilities to be tourist destinations—the public, both local and national, was encouraged to visit. Chiefly designed by Roland Wank, the elegant modernist architecture of TVA's early years served as well-crafted propaganda, furthering not only the altruism of the agency, but also that of the New Deal.[6] Seamlessly landscaped, visitors were guided onto these prominent reservations along smooth ribbons of road, and the dams revealed themselves at their best possible vantage point. Since TVA's goal was to shatter the status-quo throughout the Valley, presentation was a requisite feature. Professionalism, politics, and production were not enough—the agency sought to inspire. It displayed itself as the pinnacle of modernity, pushing marketing materials and propagandistic films to the populace in order to get them to buy into the initiative.[7]

Additionally, the presentation of the dams had to be a spectacle. They had to impound water and generate electricity in a way that moved the public, appeased Congress, and encouraged economic growth. In her essay "Redefining Landscape," Jane Wolff discusses the need for such public propaganda: "The places that TVA created argued for its agenda. Meant to look as if they had always existed, they declared that progress was not at odds with tradition, that old images and new technologies could be integrated, and that history and geography were continuous and coherent."[8] Indeed, the dam reservations fit seamlessly into the surrounding topography with minimal visual interruption and maximum public access.

As a government-owned agency, the dam reservations are public land. The roads remain unobstructed and the production of hydroelectric power is on full display. Many of the dams can be crossed by foot or automobile as well as approached by boat. Visitation is welcome and expected, and TVA operates four separate visitor centers during the recreational season. The centers are staffed by TVA retirees, and contain educational displays discussing the agency's history and importance throughout the region as well as narratives specific to each respective facility. The exhibits are well designed and informative, utilizing archival photographs, first-person accounts, and video footage. The displays contain artifacts from dam construction, and multi-media portals allow visitors to retrieve information on any TVA facility. Yet the most valuable service at these visitor centers are the TVA retirees who act as ambassadors.

I have had amazing conversations about TVA, its mission, and its history with those individuals who chose to give back to the organization for which they so proudly worked, in many cases, for decades. In my experience, their knowledge of the dams and their history is superb, and because many of them served as engineers, they welcome detailed questions about capacity, safety, and turbine variety. They also do not shy away from controversial subjects, because while they believe in the social mission of TVA and the capacity of the federal government to positively affect society, they are or once were residents of the Valley. They share personal opinions on the darker chapters of TVA's history as well as its more recent failures, such as the devastating coal ash spill at Kingston Steam Plant in 2008.[9]

Such a genuine practice of outreach does not alleviate the need for security at the dams, however. When access is denied, off-limits locations within each dam reservation are usually surrounded by chain-link fence. Additionally, the powerhouses, which were formerly open to the public, are now permanently closed. In many cases, the dam reservations have not been altered, so sidewalks and staircases that lead to the powerhouses' once public entrances now end abruptly at a barricade. Such physical blockage is minimal, visually, but it does offer an ironic juxtaposition: Architecture that was originally designed to be inviting is now inaccessible. This type of juxtaposition, when photographed in situ, testifies to the visual complexity of landscapes that not only serve multiple purposes but have served these multiple purposes over time and throughout historical, national, and global political forces—such as 9/11 and other acts of terrorism.

The central narrative of *Dangerous Waters* is not about the architecture itself, but about the places that have been created in relation to the structures' functions. *Dangerous Waters* is a visual rendering of the public spaces surrounding TVA's hydroelectric program, and it is my hope that the stories these images elicit will contribute to the ongoing dialogue about all such places. While I visited every dam within TVA's system, this book does not contain a photograph from each facility. Instead, it is a mosaic of recreational and industrial landscapes. It casts a wide representational net and includes imagery from across TVA's service area, from the smallest flood control embankments to the largest main river dams. It also contains photographs from locations that are uniquely bound to the agency through consequence, such as state and national parks that exist on land acquired by TVA through eminent domain.

Dangerous Waters most certainly is a multi-layered story and, at its heart, is about the people who live, work, and play at these TVA locations. Yet many of the photographs are absent of people, which was not by choice. The majority of the locations I visited are remote and seldom visited by anyone other than locals. Even the main river dams are frequently abandoned. It was therefore not uncommon for me to spend a few hours at some of these reservations, on a Saturday afternoon in the middle of summer, and see no more than a dozen people. At other locations, the parking lots and recreational facilities were close to overflowing. Either way, I photographed these locations and the people in them, or absent from them, as I found them.

I did not seek to gain entrance anywhere that was closed to the public, nor did I scale fences and barriers: I respected the physical and social boundaries that any citizen must. The images reveal what I was permitted to see and experience, as well as what I was not. There is an ambiguity in many of these photographs that reflects, I believe, the complexity of TVA projects from their inception to their current standing in the local and national landscape.

Whatever a viewer takes away from these images, a full understanding of these locations is found somewhere between what is expected and what it shown. Competing narratives of socialized government, architecture and utility, ecology and economics, and the social sacrifice of progress can be teased out, yet these images do not proclaim a firm stance on any of these issues. TVA is too complex for that. My personal bias toward an organization with such a staunch mission of social welfare colors these photographs as does my empathy toward what was sacrificed.

The structure of this book mirrors the geography that informs it: the Tennessee River, the watershed that drains into it, and the TVA service area that it powers. Beginning with the Clinch River, Sections One through Three move through each of the main tributary systems. Section One includes the Clinch River, Holston River watershed, and French Broad River. Continuing in a clockwise fashion, Section Two follows the Little Tennessee River and the Hiwassee River basin while Section Three explores the tributary and flood control dams found in the central and western portions of the service area. Finally, Section Four concerns the Tennessee River proper and its hydroelectric facilities. Each section contains narrative on several photographs, including historical context, if necessary, and descriptions of my time within those locations. While I do not explicitly discuss every photograph contained in this book, brief captions for every image can be found in the Appendix.

Overall, my experience traveling throughout TVA's service area was intensely emotional. I visited many of the dams on multiple occasions, and photographed their respective reservations in a variety of seasons, sometimes years apart. It was a joy to explore these locations and familiarize myself with the particular attributes and personalities of each dam.

Some locations have become as close and familiar as old friends, and I know that no matter how many times I revisit them, I will feel the same mixture of awe and sadness every time I step out of my car. I feel the weight of history in these places and understand that I am bearing witness to the success of socialized industry as well as the breakdown of certain social fabrics. Regardless of one's individual perception, TVA vistas remain landscapes of memory, loss, social progress, and technological advancement. They are visual reminders of eminent domain, government authority, and increasing matters of "security"; but, at their best, they continue to be public spaces for recreation, family, and lazy Sunday afternoons. No matter how they are perceived, one thing is certain: The dams, and the landscapes they inhabit, were built for the greater good of the Tennessee Valley and its inhabitants then and now.

NOTES

1. D. W. Meinig, "The Beholding Eye: Ten Versions of the Same Scene," in *The Interpretation of Ordinary Landscapes: Geographic Essays*, ed. D. W. Meinig (New York: Oxford University Press, 1979), 33–48.
2. The motto "Built for the people of the United States" is inscribed on the walls of many of the agency's dams and powerhouses.
3. Tennessee Valley Authority Act, 16 U.S.C. § 831 (1933).

4. "TVA Land Policy," *Tennessee Valley Authority*, https://www.tva.com/Environment/Environmental-Stewardship/Land-Management/TVA-Land-Policy.

5. "Relocated Cemeteries," *Tennessee Valley Authority*, https://www.tva.com/Environment/Environmental-Stewardship/Land-Management/Cultural-%2B-Historic-Preservation/Relocated-Cemeteries.

6. Jane Wolf, "Redefining Landscape," in *The Tennessee Valley Authority: Design and Persuasion*, ed. Tim Culvahouse (New York: Princeton Architectural Press, 2007), 26–27.

7. Steven Heller, "TVA Graphics: A Language of Power," in *The Tennessee Valley Authority: Design and Persuasion*, ed. Tim Culvahouse (New York: Princeton Architectural Press, 2007), 105–6.

8. Jane Wolf, "Redefining Landscape," in *Tennessee Valley Authority*, 52.

9. Stephanie Smith, "Months after ash spill, Tennessee town still choking," *CNN*, last modified July 13, 2009, http://www.cnn.com/2009/HEALTH/07/13/coal.ash.illnesses/index.html?eref=rssus.

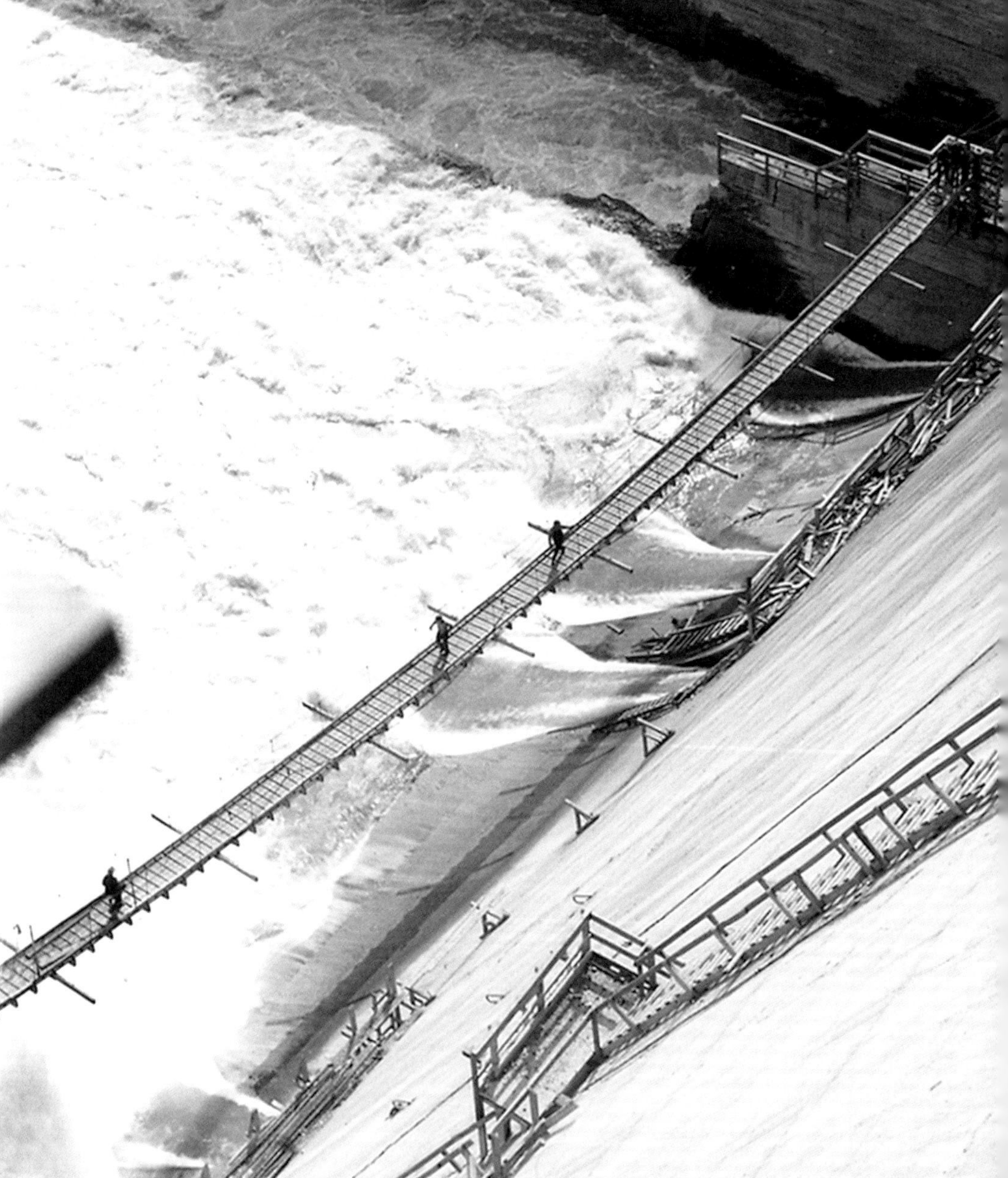

Norris Dam (1936) / Tennessee Valley Authority

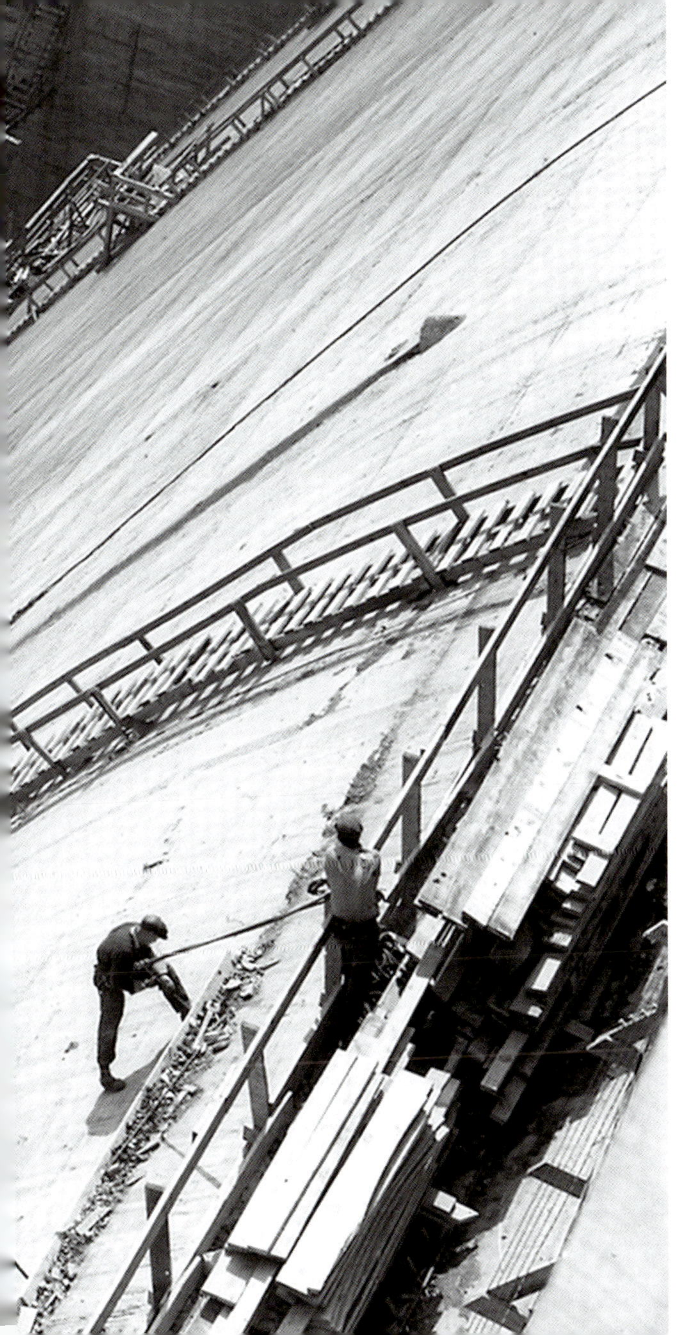

SECTION ONE

Northeastern Tributaries

Clinch River

Dangerous Waters begins appropriately with Norris Dam as it was the first hydroelectric facility the Tennessee Valley Authority built. *Norris* (19) was also the first photograph I made for this project even though I had spent a week visiting other dams, hiking through their reservations, and visiting their relocated cemeteries.

In that first week, I had been having trouble experiencing TVA spaces in a way that meshed with my initial desire to move beyond photographing a landscape simply for its history and iconic architecture. It was not until I was standing on the top of Norris Dam on a hot June morning in 2013 that I realized the social narrative of these locations lies within the nexus of architecture and its utilization. When I leaned over the edge of the dam, I perceived for the first time that these structures were not simply representations of their past but were also metaphors for their social aims—past and present, intended and not intended.

What I experienced was as beautiful as it was unsettling: The sounds generated by the dam's two turbines echoed in my ears and up the valley, and my body vibrated along with the concrete. I was standing on an operating hydroelectric dam that is considered by many to be one of the most storied and striking in the country, yet the experience of that place was not about the structure's history but rather what the structure was enabling. I turned around and looked out over Norris Lake. What I saw was a thriving landscape of leisure in the service of utility. For the first time in my fieldwork, instead of being distracted by historical facts or preoccupied by how to get past them, at Norris Dam I was fully present—and *Dangerous Waters* began to take shape. *Norris* proclaims the visual focus of this project through the imagery of concrete and water.

At a height of two hundred and sixty-five feet,[1] Norris Dam does not overwhelm visitors with intensity or grandeur but wins them over with charm. As the first major construction of TVA, Norris Dam had to make a spatial and architectural statement. More than any other TVA dam, Norris symbolizes the agency's original utopian approach to regional development. The impact of its design and the ambition of TVA's planners stretched past the reservation's borders to include the adjacent town of Norris, a new state park, and a twenty-one-mile freeway. The Norris Project's enormity was fully aligned with TVA's progressive experimentation: The agency viewed its completion as the first step in improving people's lives in the Valley.

The creation of Norris Dam left a number of scars in and around the reservation. *Norris, Marina* (36) displays the sheared wall of a quarry where stone was gathered and crushed for the concrete mix used to construct the dam. Its close proximity to the structure now serves as an easy-access point for boat storage and reservoir access. The construction of Norris Dam also created Big Ridge State Park. Built in 1934, the park is situated on a sub-impoundment along the southern shore of Norris Lake. It was conceived as a model display of lake recreation meant to showcase the many outdoor opportunities now available to local families after the waters rose over their former homesteads. Currently owned and operated by the Tennessee Department of Environment and Conservation, it remains an example of

TVA's understanding that dam construction was a two-part process: reservoir preparation and reservoir utilization. It has served four generations of the dispossessed and suggests a simpler time of innocence and happiness, a space of rejuvenation that is indebted to what rests far below its waters.

Melton Hill Dam, approximately fifty-seven miles upstream, was constructed thirty years after Norris and is the only tributary dam in TVA's system that features a navigation lock. Melton Hill is nondescript: it does not have the flair of its older Clinch River sibling, and instead opts toward a more utilitarian design. While the dam's importance to river navigation and economic development is not explicitly proclaimed through its architecture, the facility extends barge transportation to nearby Clinton, Tennessee. With its minimized presence along the river, Melton Hill prompts most visitors to look past it and to embrace the surrounding mountains. *Melton Hill, Upstream* (21) contemplates this unassuming tension. The dam's flat metal industrial components contrast with the adjacent overabundant green shoreline, and its muted color is enhanced by

the overcast sky. It began to rain as I made this photograph, and small ripples caused by the raindrops
can be seen on the surface of the water.

Holston River Watershed

The Holston River watershed includes five TVA dams,
with Watauga Dam, ensconced within Cherokee National Forest, being the furthest east. Watauga Dam
impounds one of the most scenic reservoirs in TVA's
system, and despite its massive size, does not an-
nounce itself in an overt manner. Instead, it is situated back from the main recreational area, viewable
on land only from a specific stone overlook that lies a
short hike from the visitor center parking lot. Three
hundred and thirty-two feet high, its earthen-work
construction blends seamlessly into the rocks and
foliage of the adjacent mountains, and this vista is
only interrupted by two intake towers that stand on
either side of the dam.[2]

The top of Watauga is accessible only by foot,
either by following the Appalachian Trail, which
crosses over the dam, or hiking up a steep TVA access road. Looking downstream through the valley
and toward the detached powerhouse, the height
and density of Watauga Dam's construction is evident. Here visitors can fully grasp the topography
that lies beneath the reservoir's waters as the mountains rise steep from the shoreline.

The focal point of Watauga's lakefront recreation
area is a simple and attractive visitor center with a
glassed-in overlook that reflects the blues and greens
of the surrounding landscape, but as of the time of
this printing it remains closed to the public. A barely
readable educational display that discusses the relocated town of Butler can be seen through the visitor
center's front doors and seems to be an afterthought
in TVA's effort to pay homage to one of the communities it displaced. During my first visit to Watauga,
in January of 2014, I watched a middle-aged couple
read the welcome sign and recreational map found
in the nearby parking lot. The woman then walked
toward the visitor center with interest and pulled
on the door. Not expecting it to be locked, she pulled
again and turned to her husband.

"It's locked," she said.

Her husband gave an inquisitive look and called back, "What do you mean it's locked?"

She put her hands in her pockets and returned to his side. "I can't get in. I guess this sign is all they have."

Such a frustrated interaction with public space at TVA's hydroelectric reservations is not a rare occurrence. Many of the dams completed in the years after World War II possess detached visitor centers that have been, until recently, closed to the public due to a variety of issues including funding and security. In true TVA fashion, they are purposefully designed to be inviting and architecturally striking, with plate glass windows and benches of polished wood. There are four within the Holston River tributary system including the one at Watauga. South Holston's visitor center stands on a road that dead-ends at a small picnic area near the the top of its dam. Fort Patrick Henry's visitor center is a more nondescript brick building set along a landscaped circular drive near the dam's powerhouse. Finally, Boone Dam's visitor center sits prominently at the top of a hill and overlooks extensive recreational facilities. Both South Holston and Fort Patrick Henry's

visitor centers were refurbished and reopened in July 2015, demonstrating TVA's commitment to public access. The photographs I made of those two facilities during my fieldwork remain as components of this project; however, I have also included an image of South Holston's in its refurbished state.

My time at Boone Dam in June of 2013 illuminated the contemporary relevance of TVA's hydroelectric reservations and the communal opportunities they provide. I witnessed hundreds of people engaging with the landscape and using TVA-managed amenities within designated recreation zones, all of which were adjacent to active hydroelectric power stations. This singular space provided both electricity for the community and a location for social communion and personal relaxation. *Boone, Recreation Area* (23) illustrates the interaction of landscape, utility, and community while contemplating the visual and cultural compromises we make as a society for the necessity of electricity. The image was captured on a Saturday afternoon when the location was near-capacity. Every picnic table and pavilion was occupied, the overflow parking was close to full, and the swimming area was bustling with sunbathers and children. While these amenities are free to local citizens, there is still a price: the gray band of rock stretching into the water from the right of the frame. While the dam might be out of sight, it is never truly out of mind.

Many photographs in this book exhibit the multipurpose utilization of landscape seen in *Boone, Recreation Area*. TVA benefits the people of the Tennessee Valley by placing the interests of the many over the few, and I have come to see in the course of my fieldwork that this mission is not one of disingenuous corporate altruism but is in fact a sincere guiding principle.

While these hydroelectric locations are complex, both visually and culturally, the perception of them today is absolutely colored by the past. Many families have passed down stories of both forced removal and economic opportunity, and many have voiced displeasure when unused acquired land was given to other government agencies[3] or auctioned off to speculative developers.[4] In this regard, a dam that was built in the early 1950s, such as Boone, must continue to positively affect the society for which it was built, both with affordable electricity and publicly

accessible communal space. Such a commitment is currently being tested at this very facility as a sinkhole near the dam revealed substantial seepage in 2014. The construction is scheduled to last until 2020, and the surrounding community must endure a significant drawdown of Boone's reservoir. While this construction is unsightly and inconvenient for the private homes that encircle Boone Lake, TVA's commitment is to the region as a whole and repair of the dam was determined to be the best option.[5]

French Broad River

While TVA controls only two dams within the French Broad River watershed, Douglas Dam, one of the most important structures to TVA's identity, is included in that small count and plays a significant role in the region. Few projects within the history of TVA possess more symbolism and mythology than Douglas Dam as the story of its construction has taken on mythic qualities within the organization.[6] Rushed into production in February of 1942, Douglas Dam was completed in a remarkable three hundred and eighty-two days. The first of TVA's World War II dams to be completed, it delivered immediate electricity for aluminum production. The completion of this dam, a labor of nationalism and necessity, proved that ingenuity and determination could overcome long odds. It showcased the organizational power of TVA and evangelized its social message: to provide for the greater good through the resources and people of the Valley.

One look at its well-maintained visitor facilities that includes a sleek open-air structure set atop a hill overlooking the dam and its reservoir, indicates that the dam is still of great importance. However, my photograph, *Douglas, Visitor Center* (30), is meant to posit the question: "Why does this visitor center exist"? During my fieldwork to this facility, I saw a total of fifteen people, which suggests that few people know the "myth" of Douglas and it is not popular among vacationers or locals. Positioning the camera where a tourist would not, I consciously refused to photograph the persuasive vista that this building was intended to provide.

NOTES

1. "Norris," *Tennessee Valley Authority*, https://www.tva.com/Energy/Our-Power-System/Hydroelectric/Norris-Reservoir.
2. "Watauga," *Tennessee Valley Authority*, https://www.tva.com/Energy/Our-Power-System/Hydroelectric/Watauga-Reservoir.
3. Duane Oliver, *Hazel Creek from Then Till Now* (Maryville: Stinnett Printing Company, 1989), 93–95.
4. Erwin C. Hargrove, *Prisoners of Myth* (Knoxville: University of Tennessee Press, 2001), 169–71, 174.
5. "Boone Dam Project," *Tennessee Valley Authority*, https://www.tva.gov/Newsroom/Boone-Dam-Project.
6. Hargrove, *Prisoners of Myth*, 56–57.

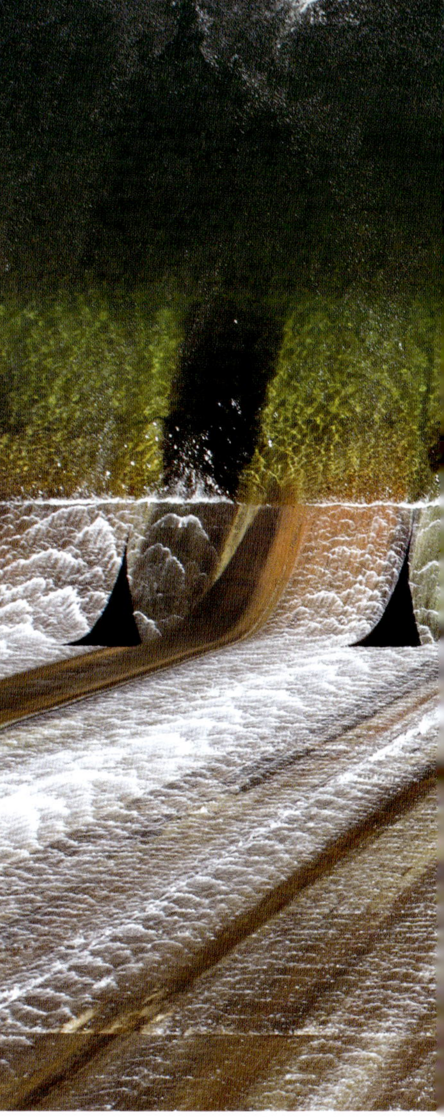

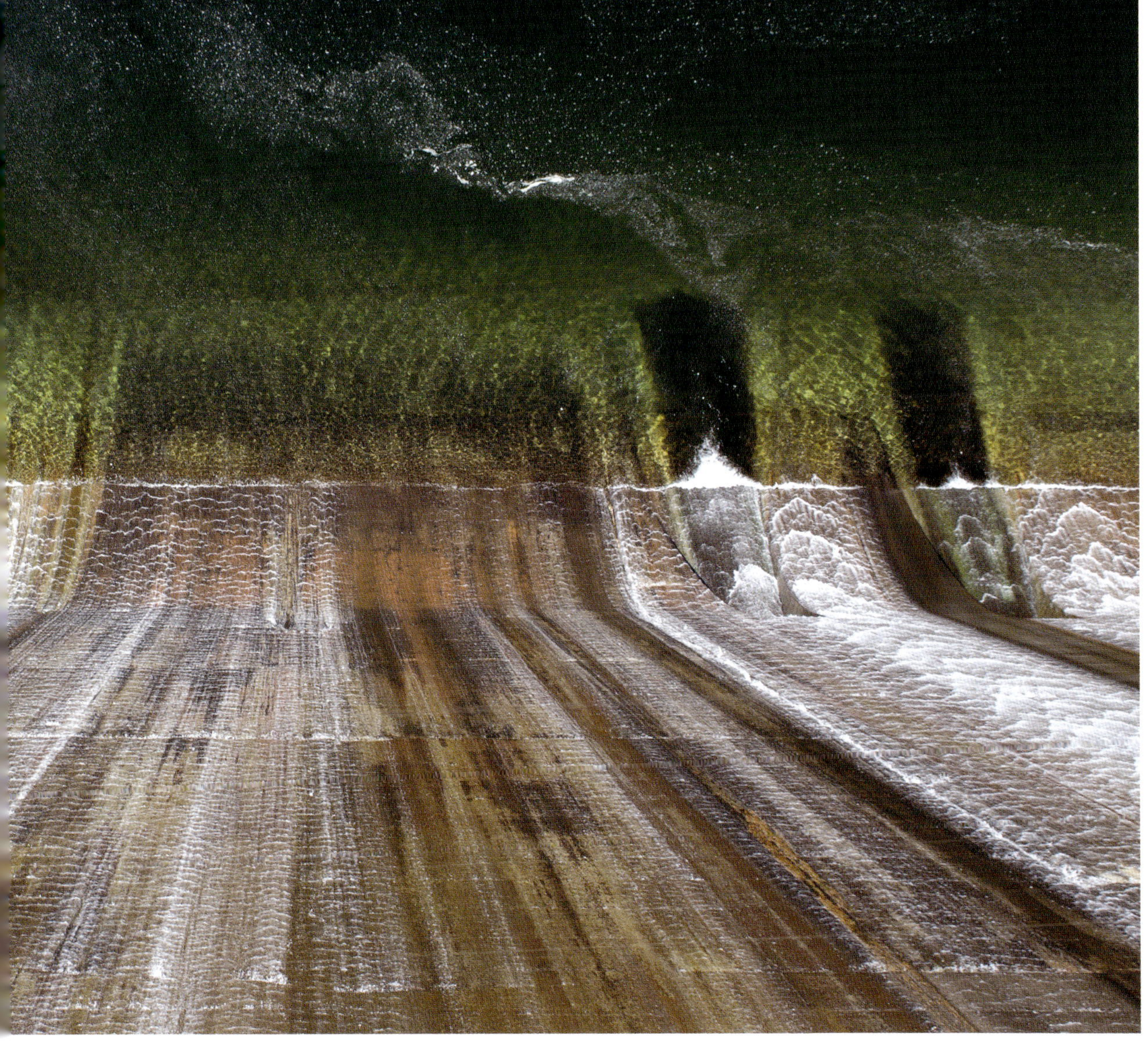

Norris

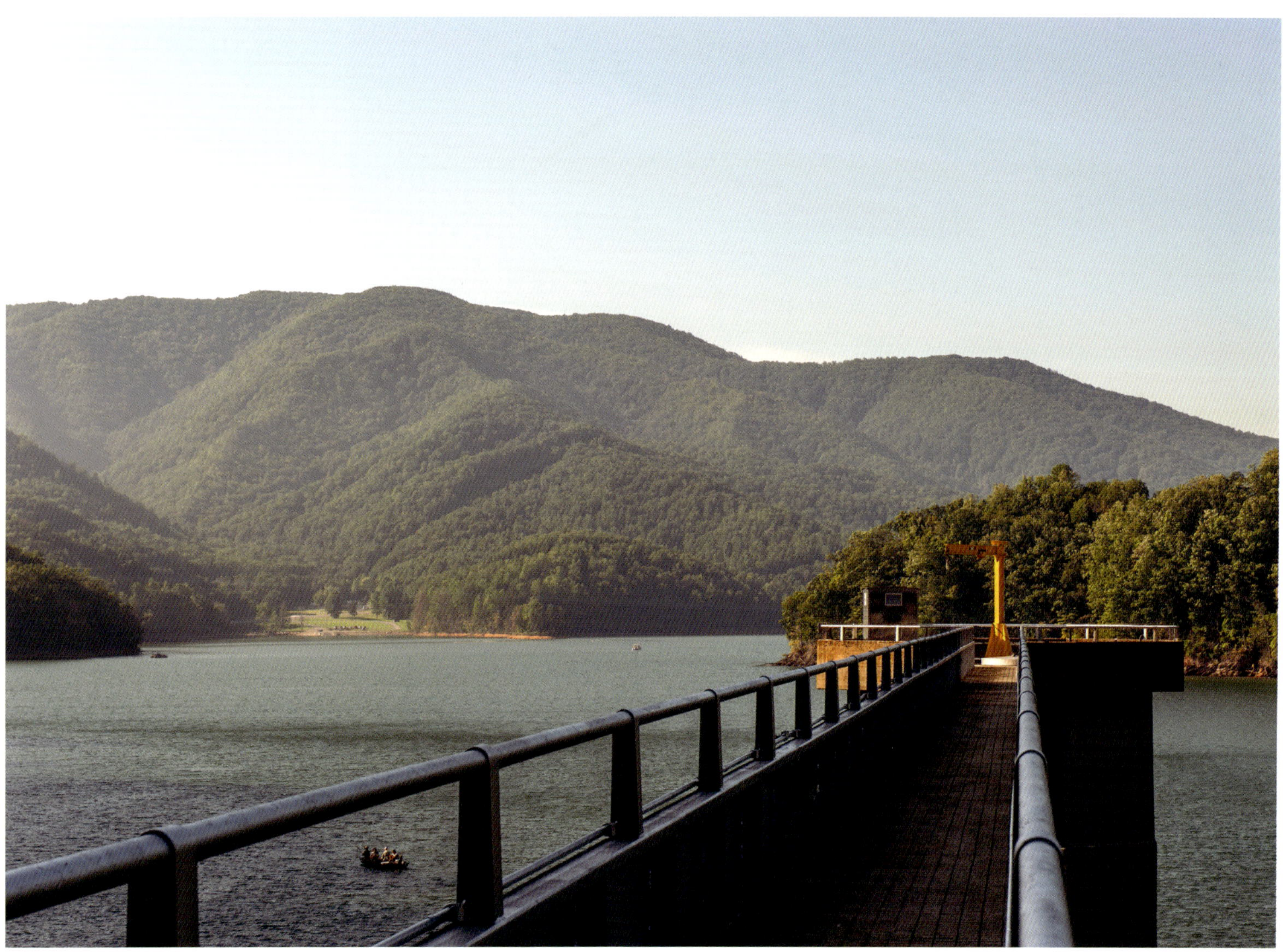

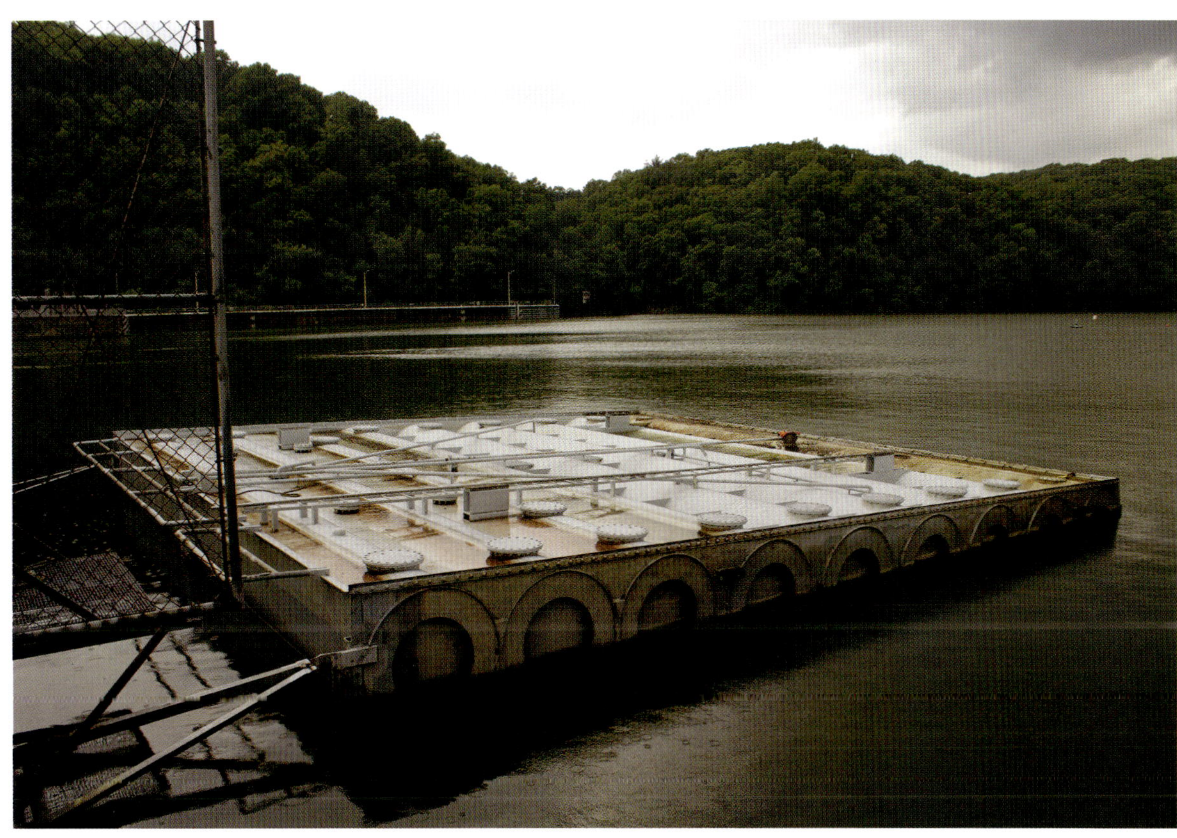

Melton Hill, Upstream

(facing) Watauga, Reservoir

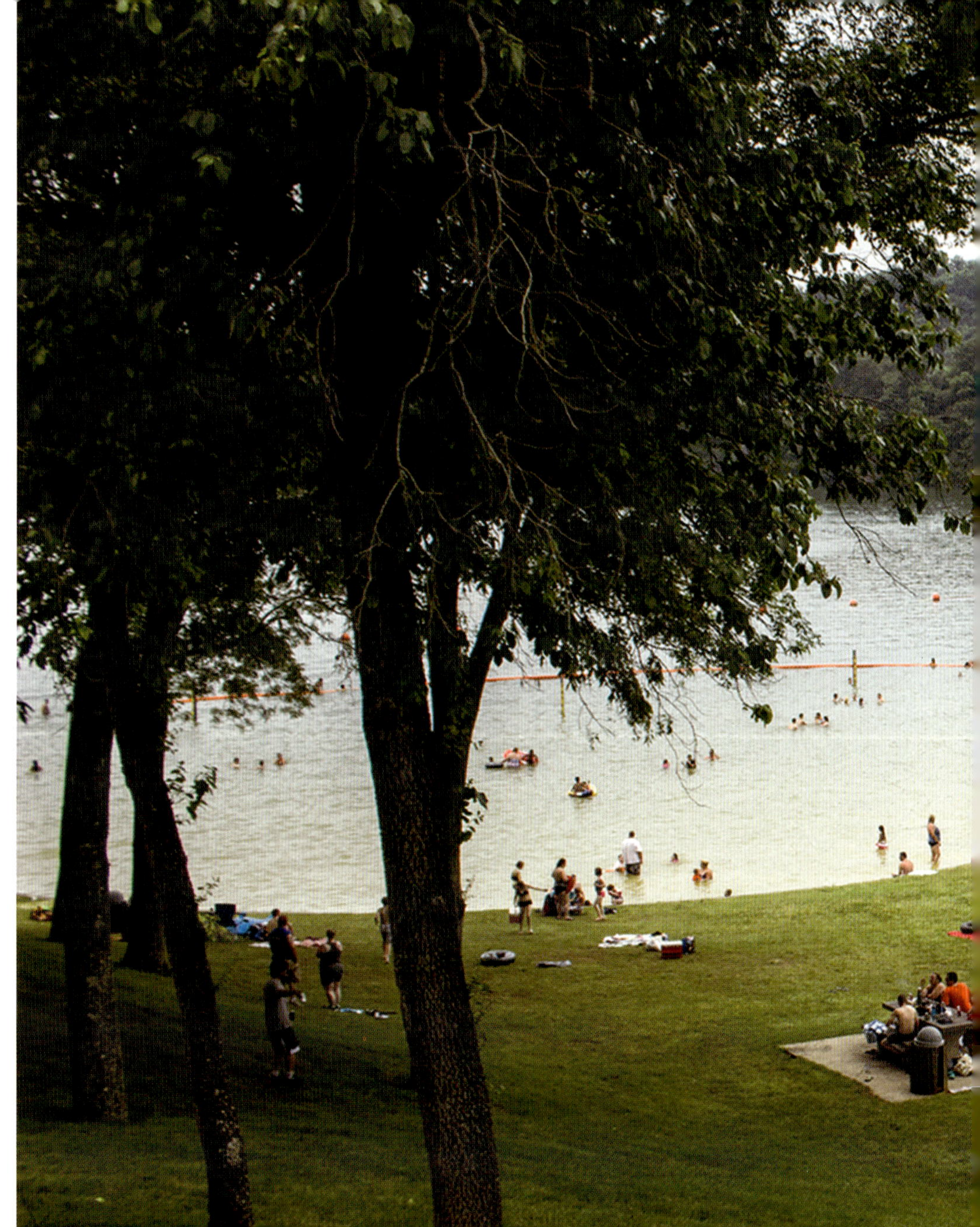

Boone, Recreation Area

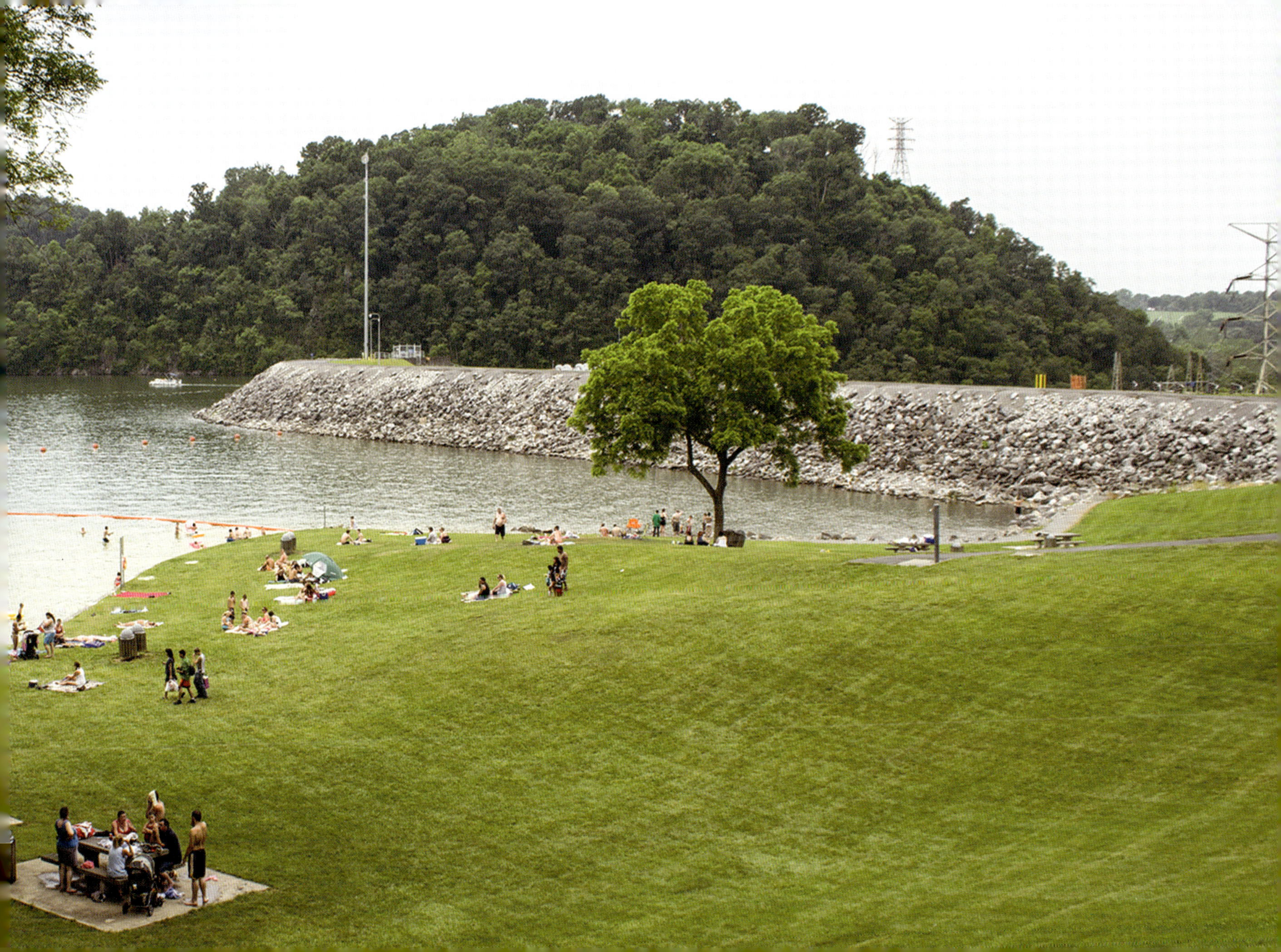

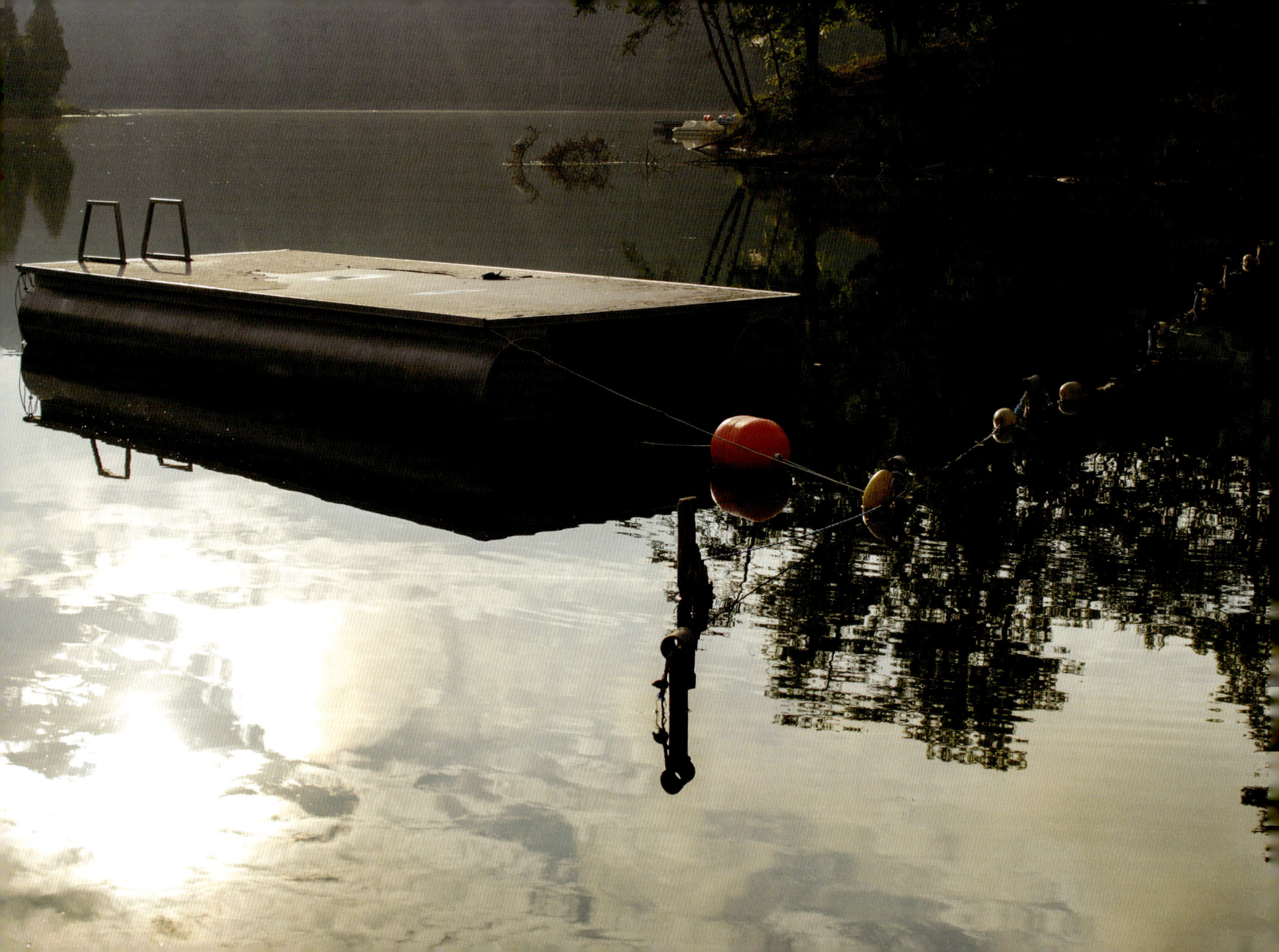

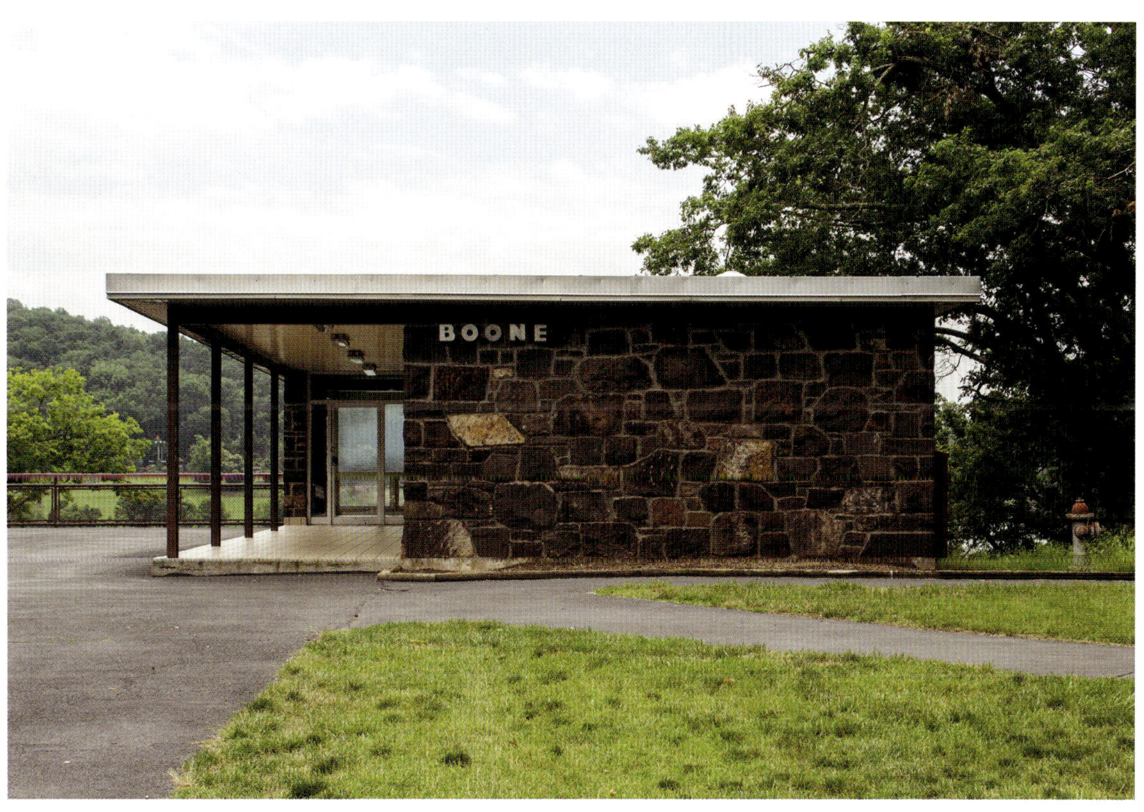

Boone, Visitor Center

*(facing) Big Ridge State Park,
Swimming Area*

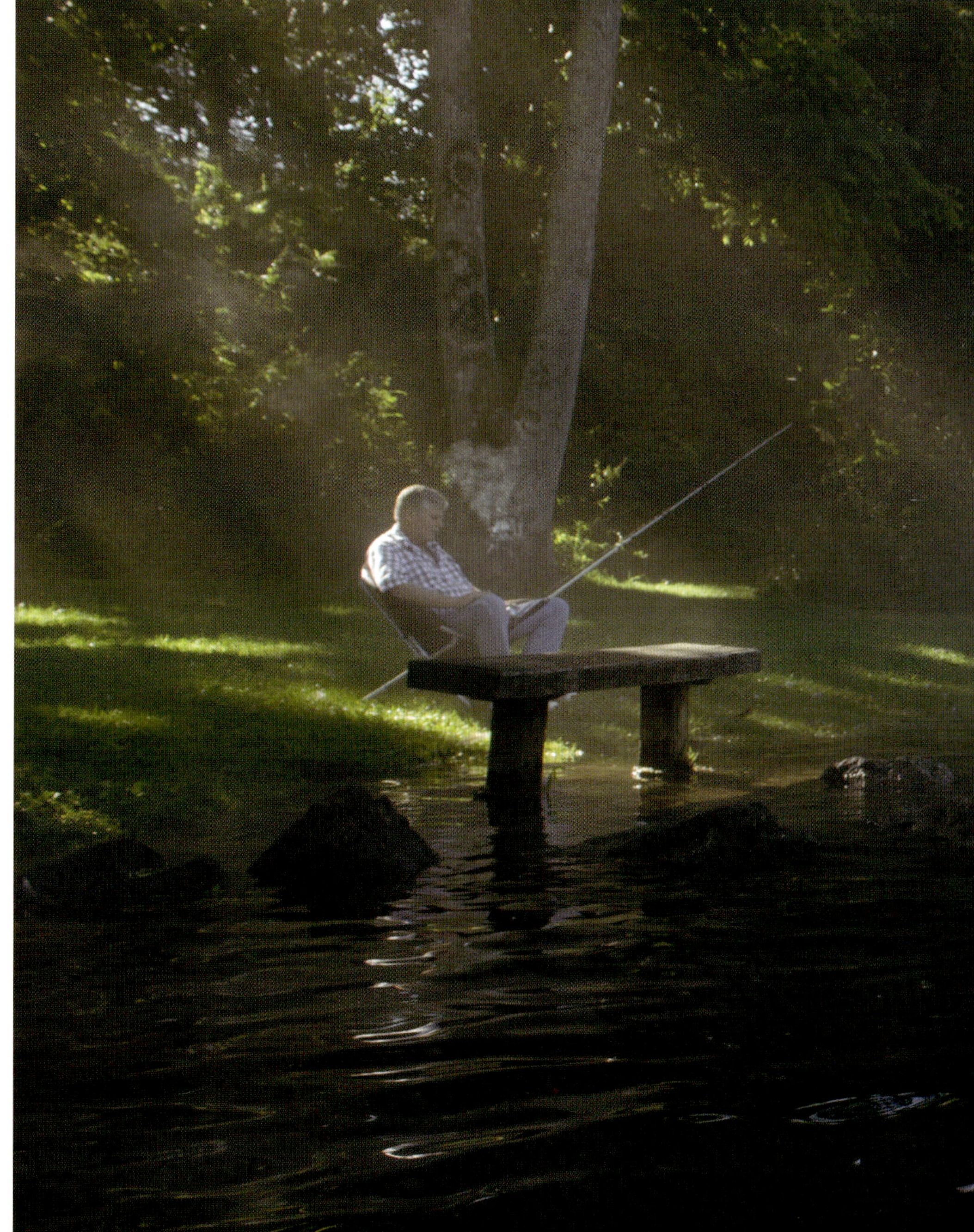

Norris, Downstream

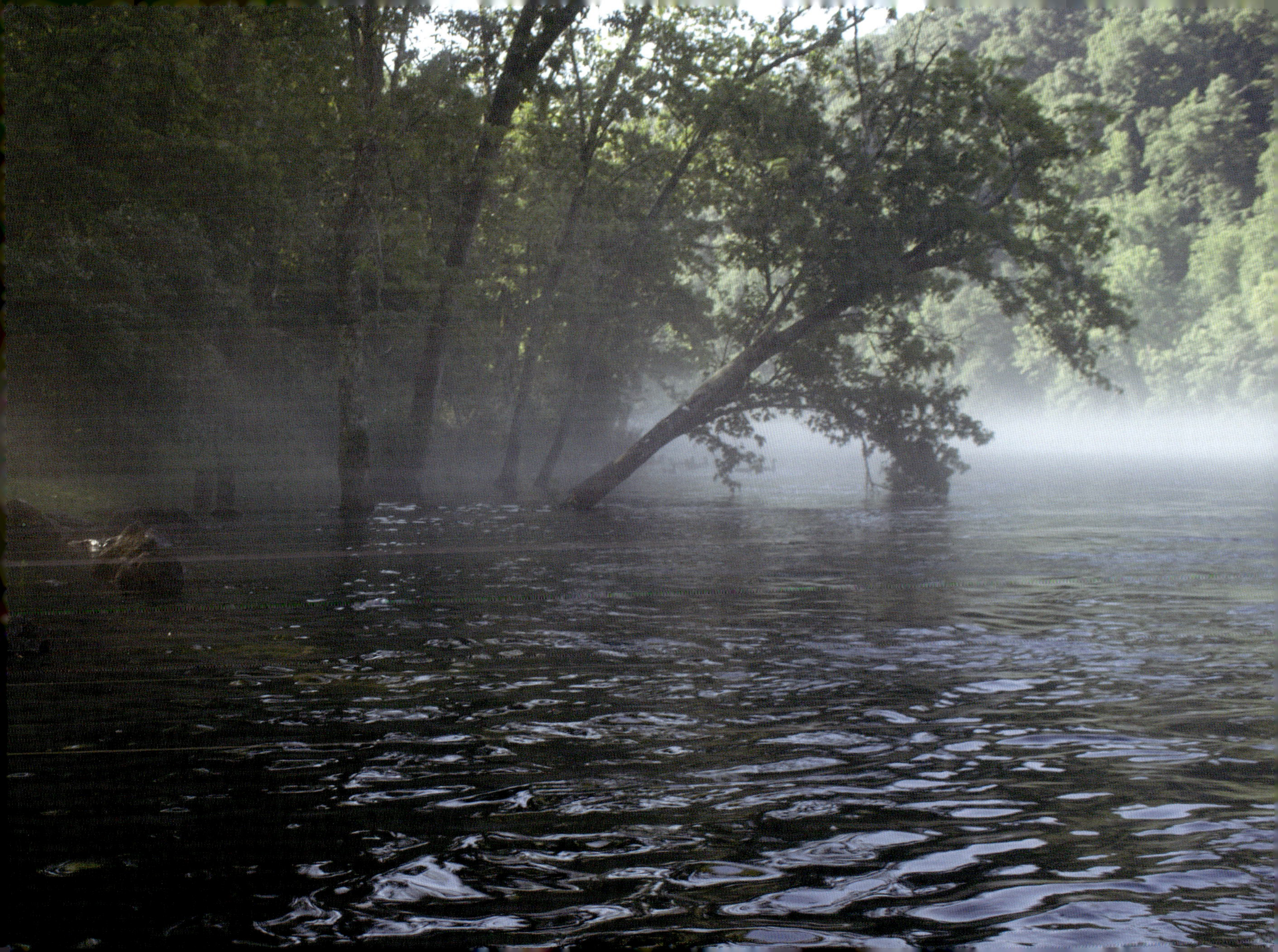

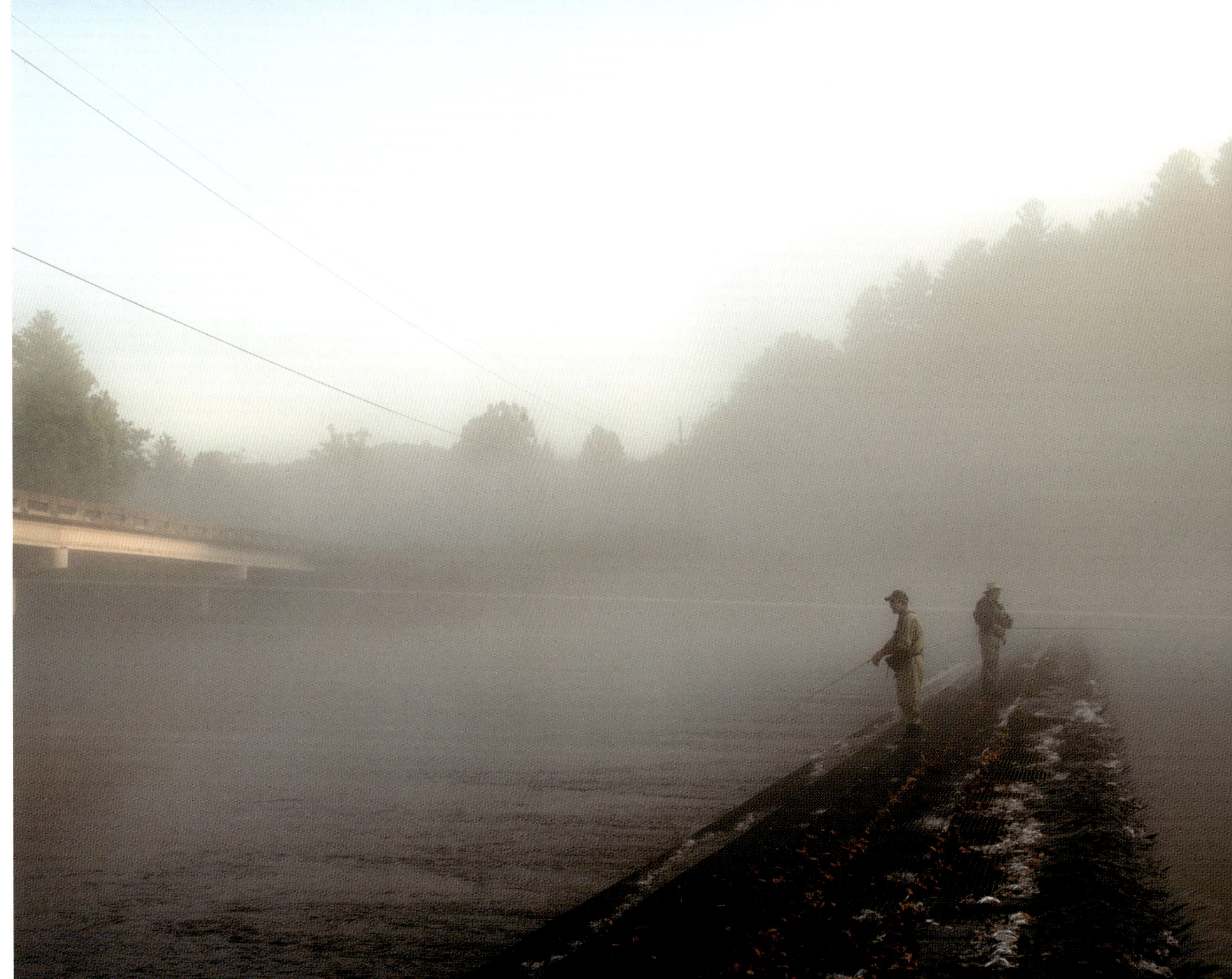

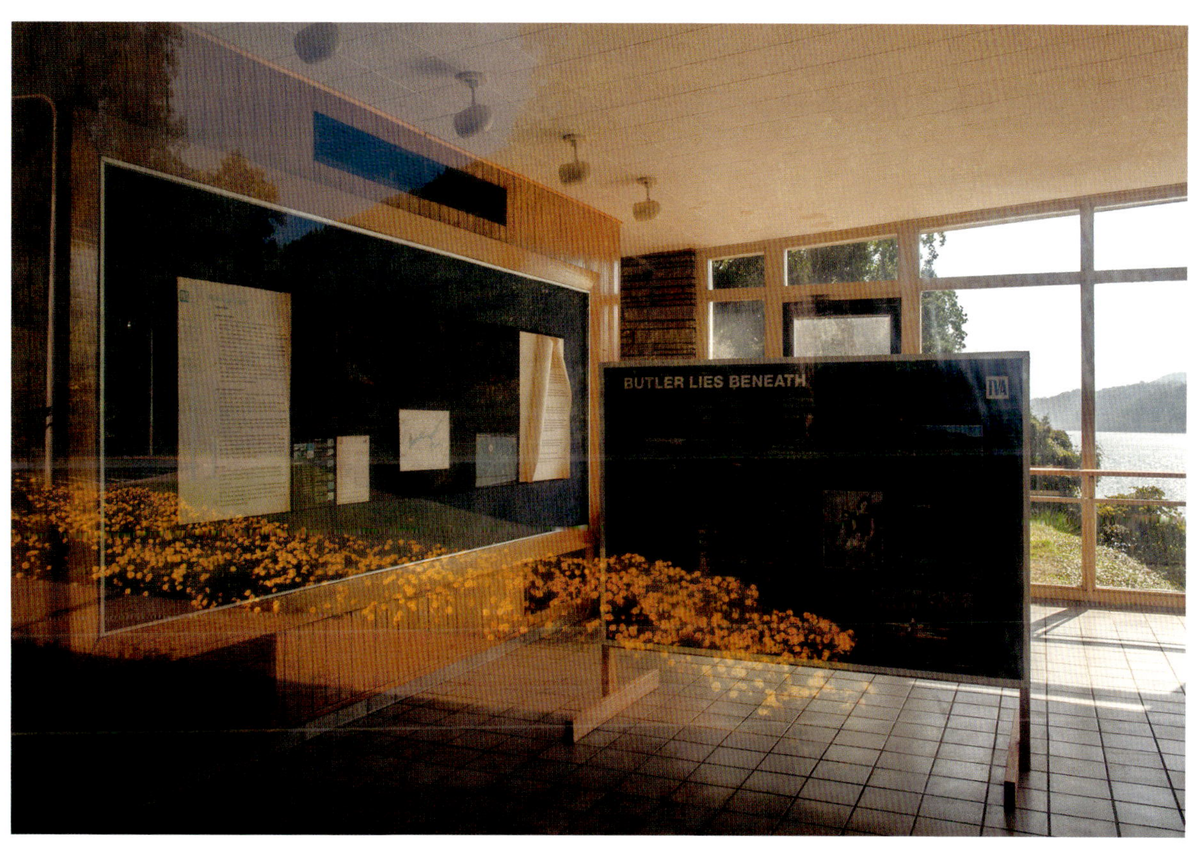

Watauga, Visitor Center

(facing) South Holston, Downstream

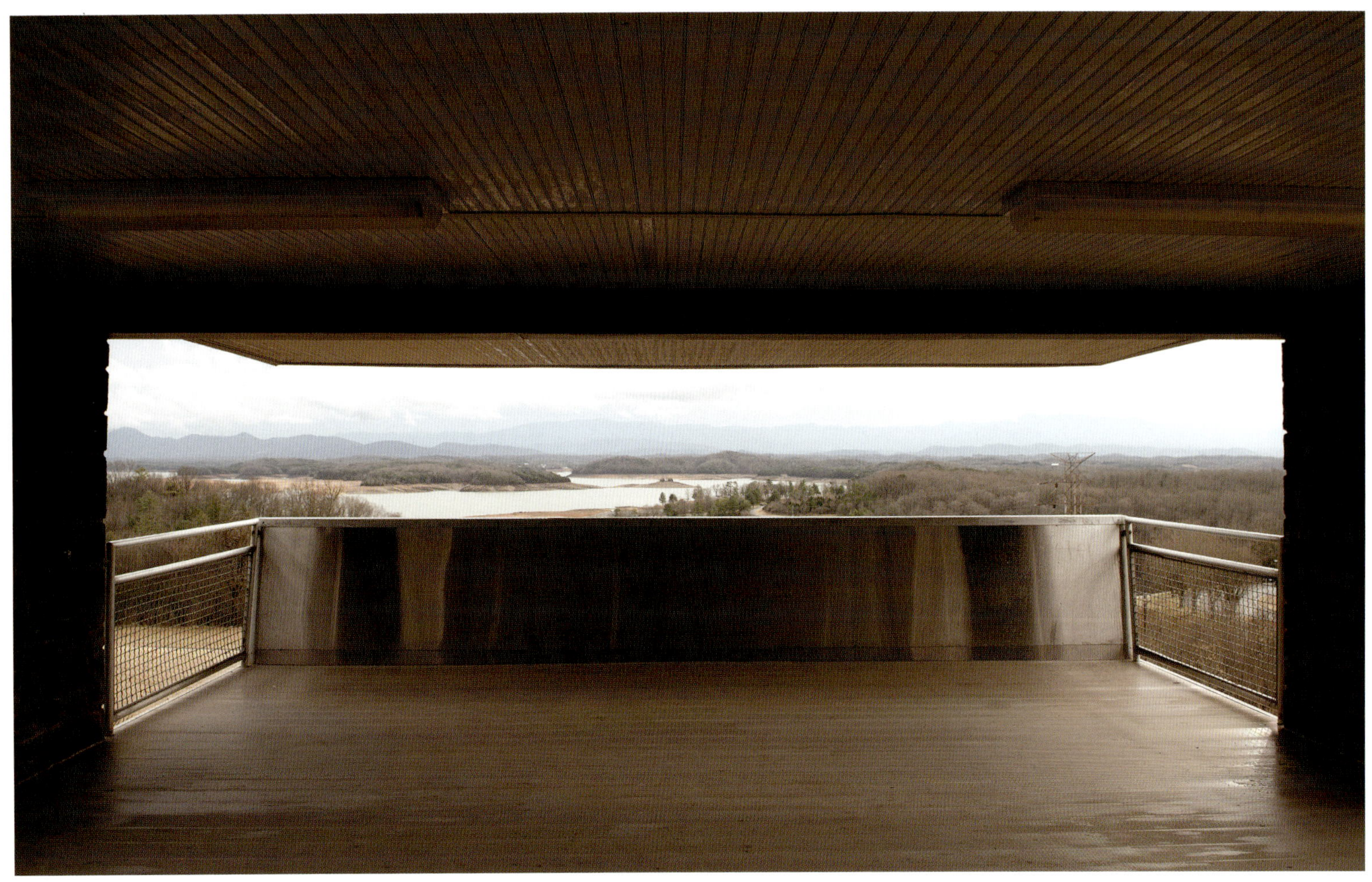

Douglas, Visitor Center

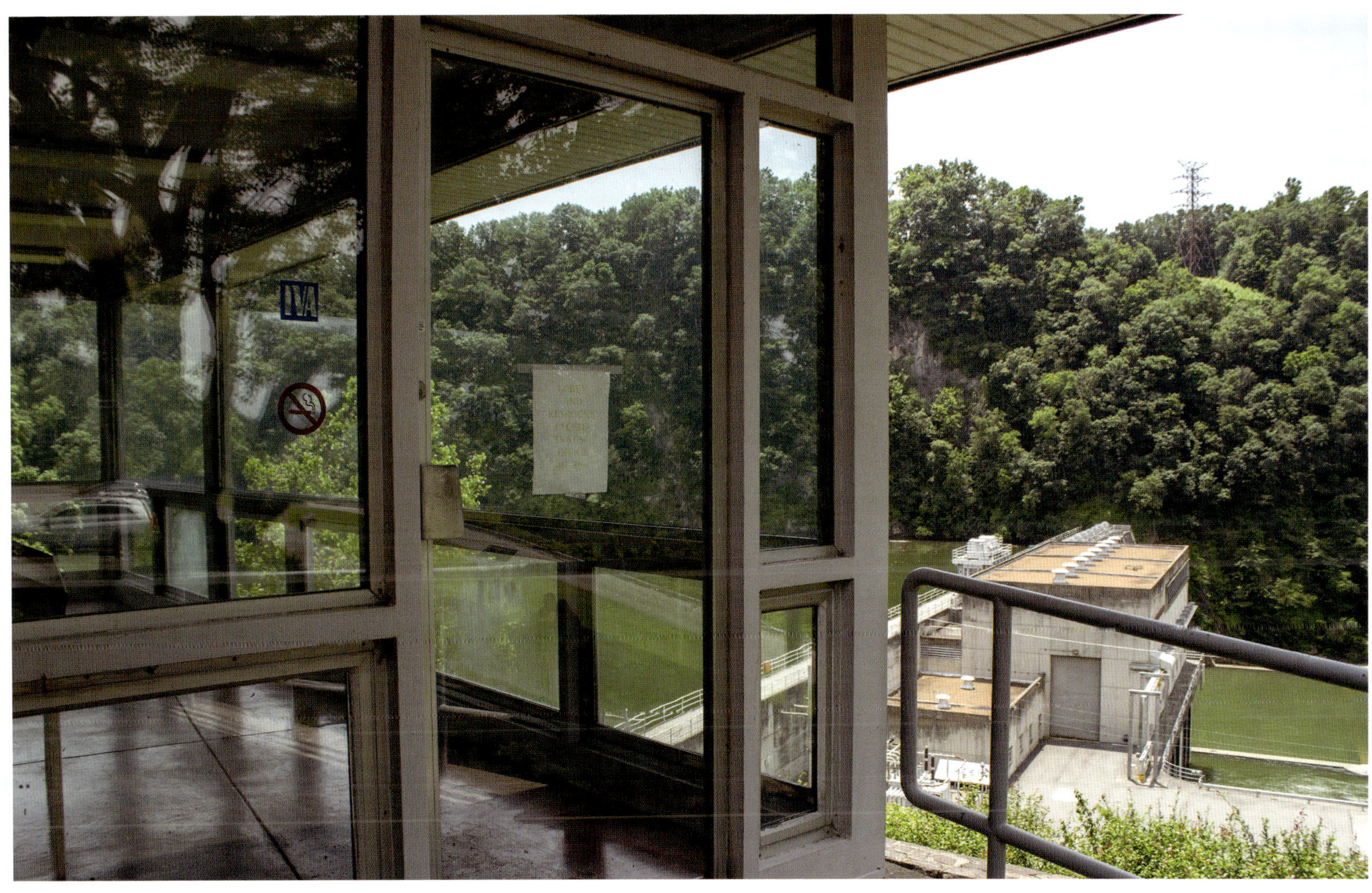

Fort Patrick Henry, Visitor Center

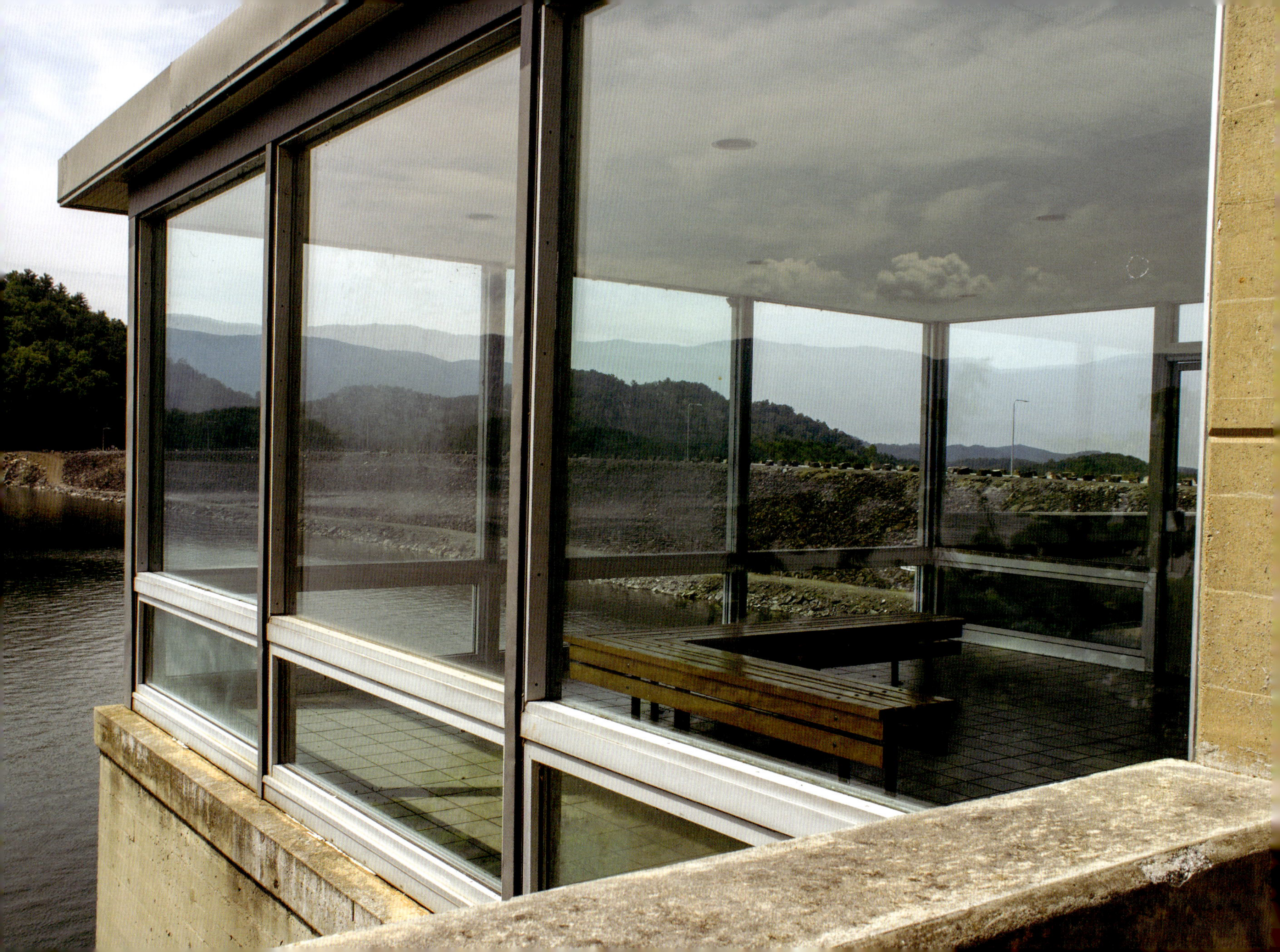

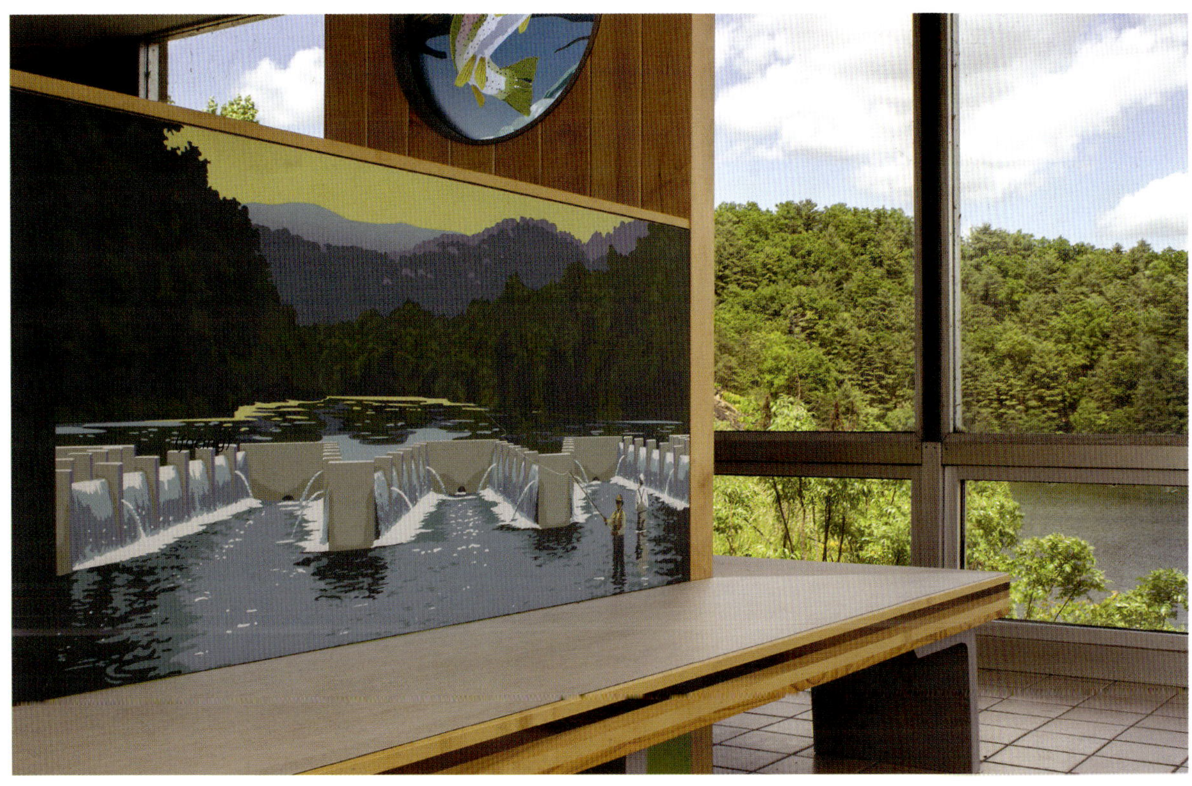

South Holston, Visitor Center (Reopened)

(facing) South Holston, Visitor Center (Closed)

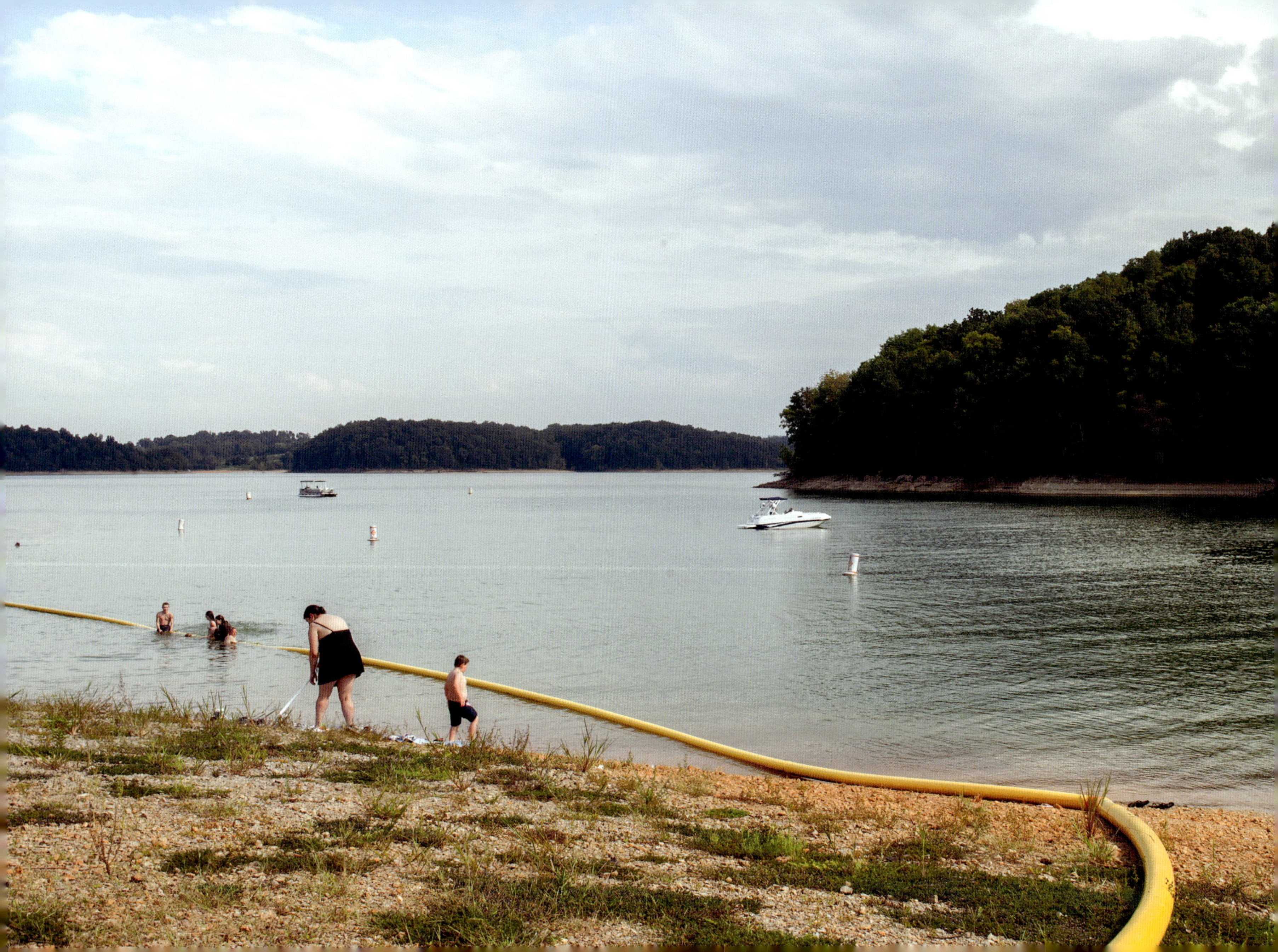

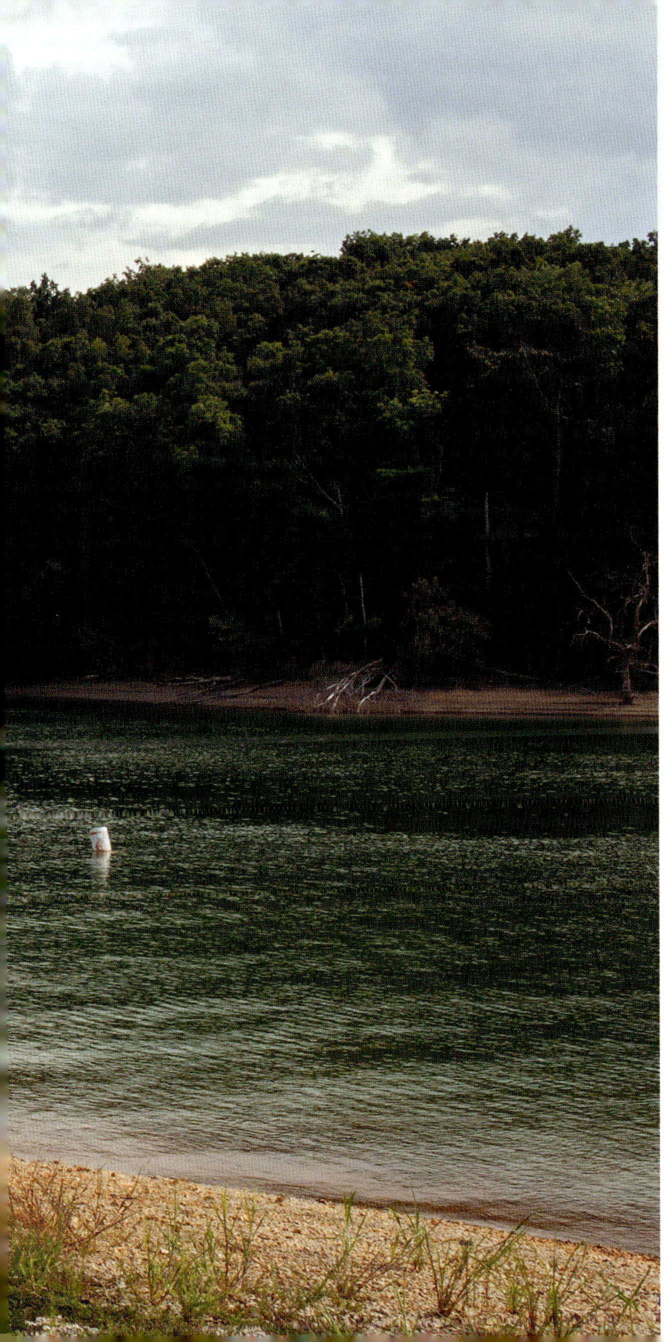

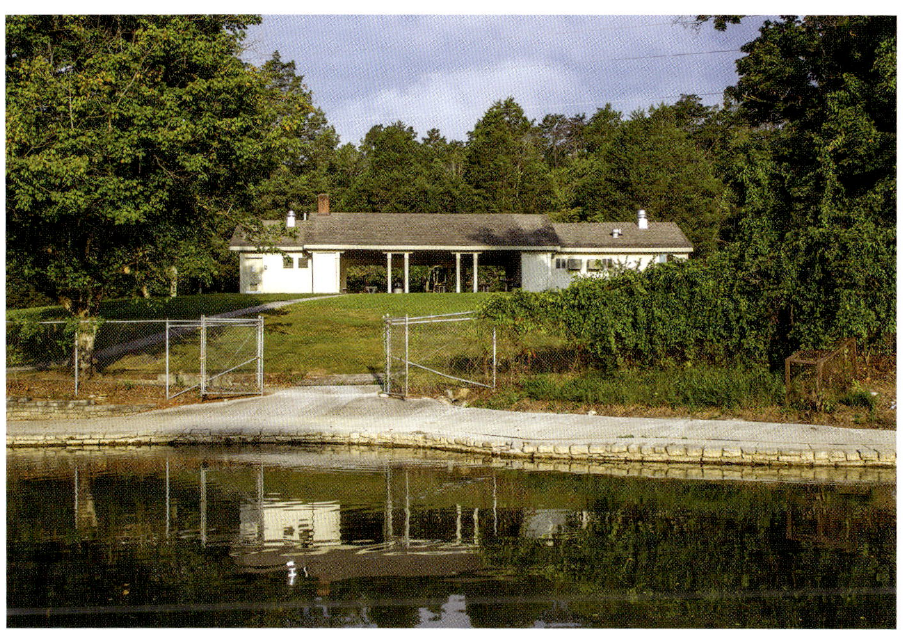

Big Ridge State Park

(facing) Cherokee, Recreation Area

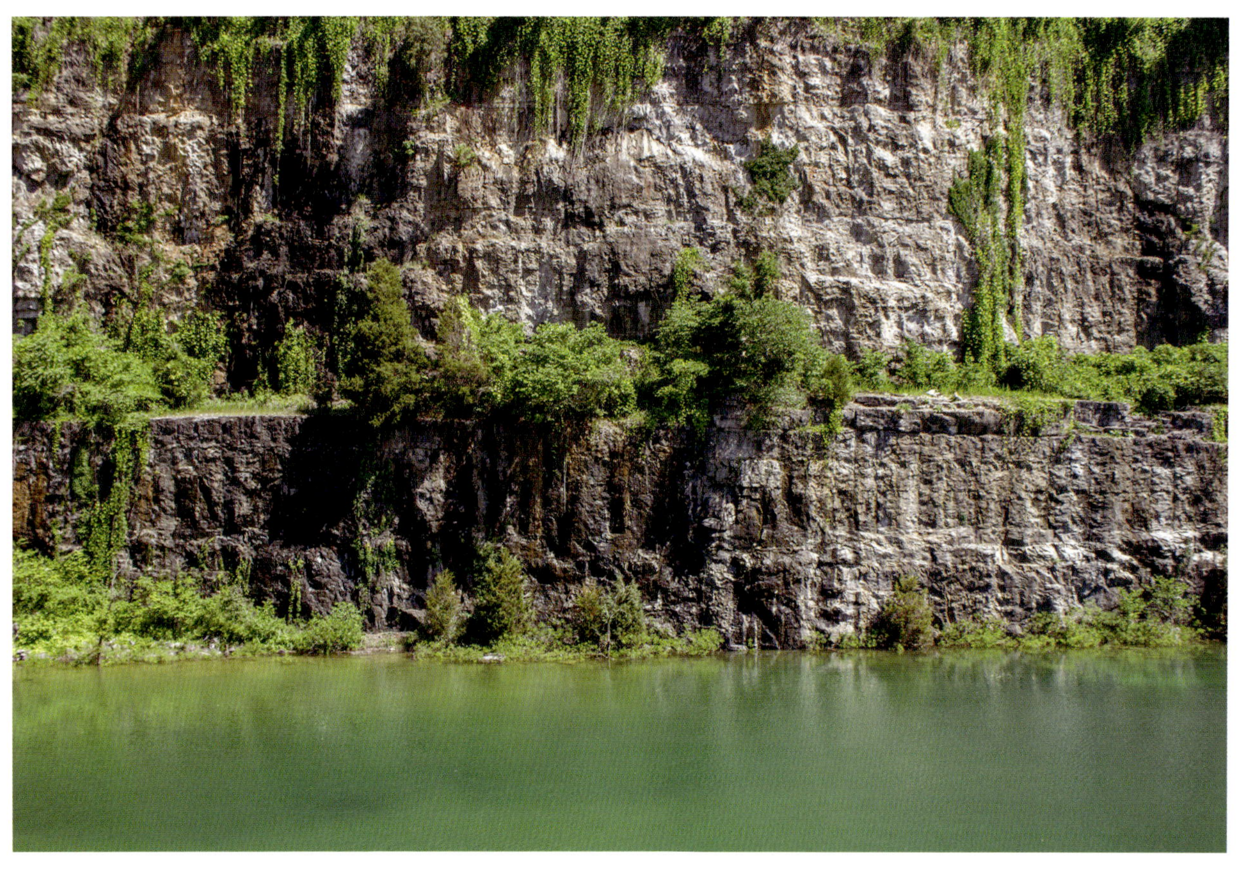

Norris, Marina

(*facing*) *Wilbur*

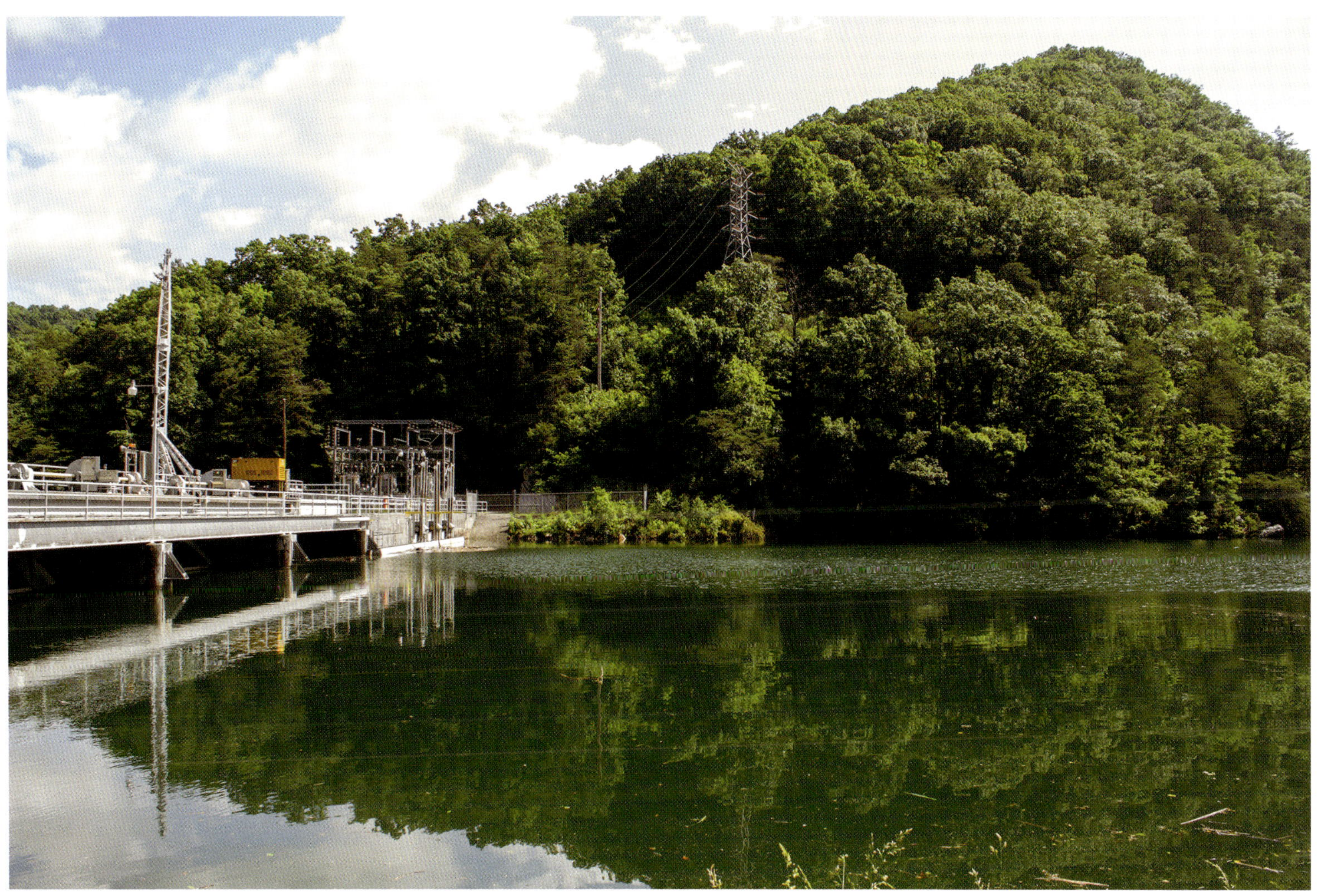

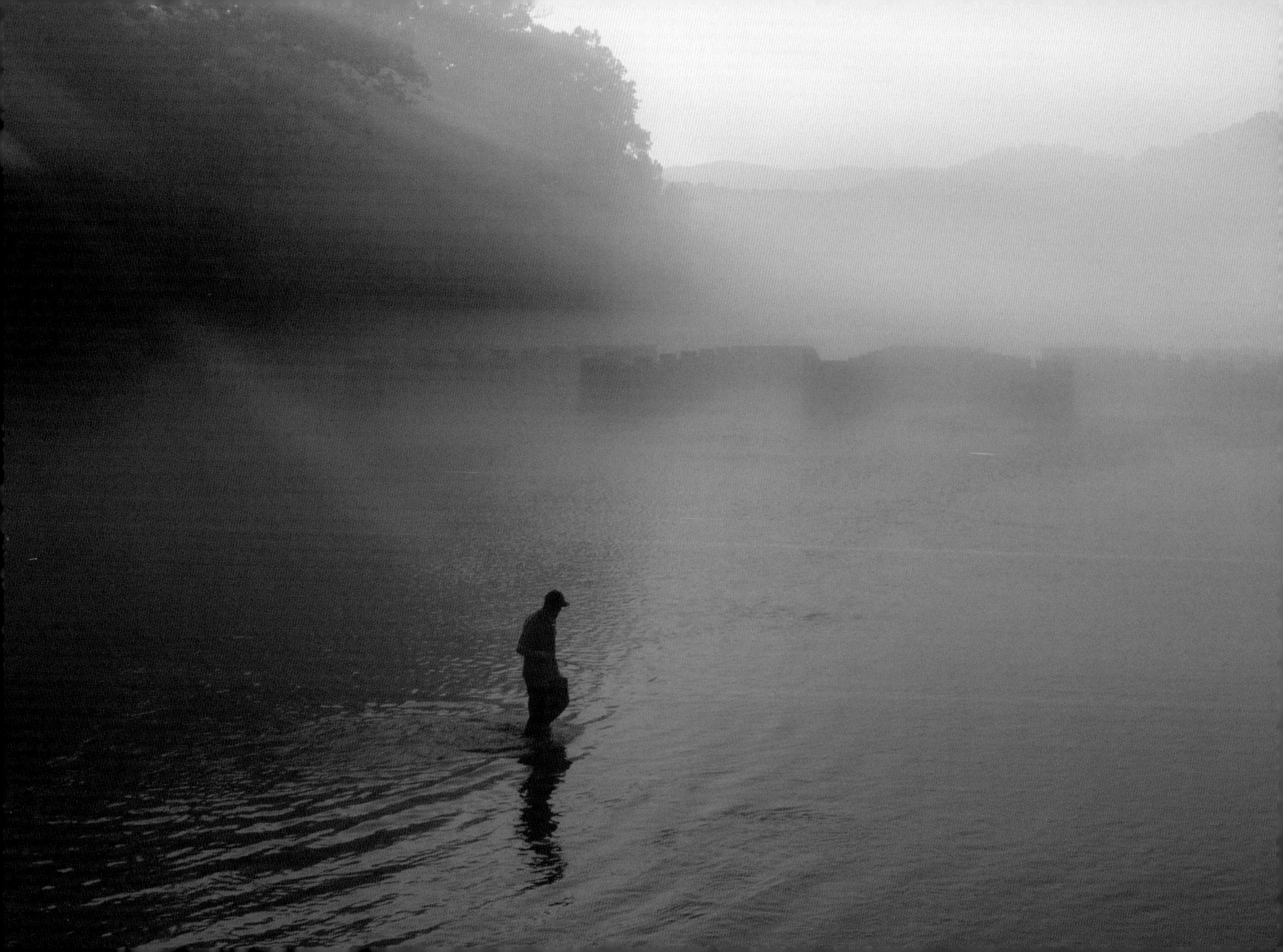

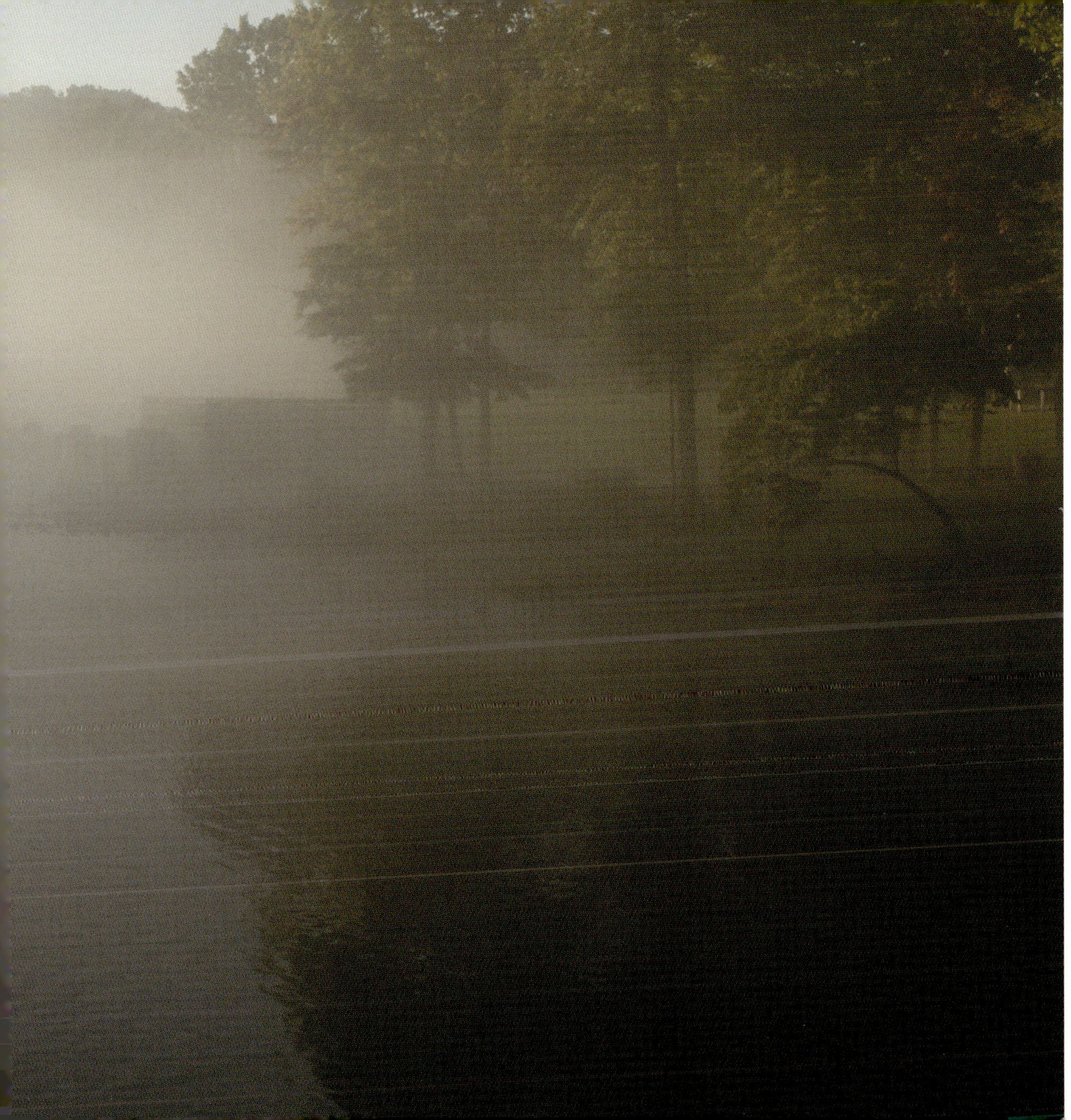

South Holston, Infuser Weir

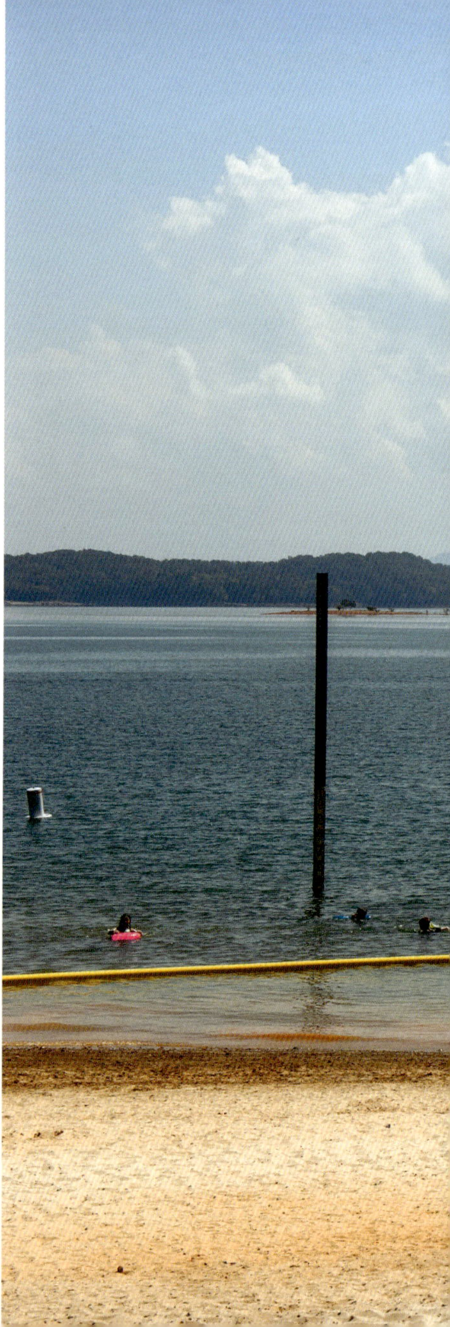

Douglas, Recreation Area

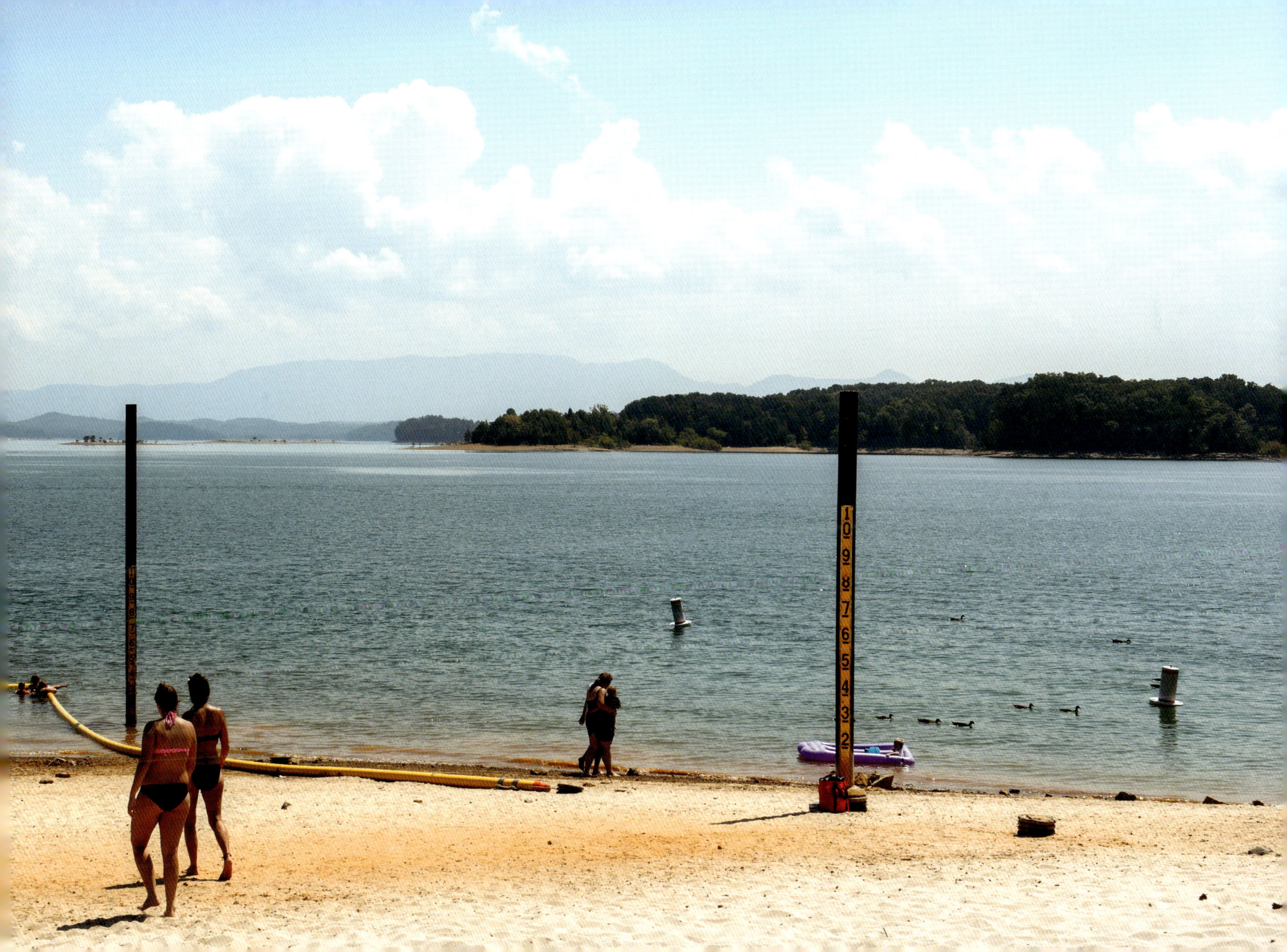

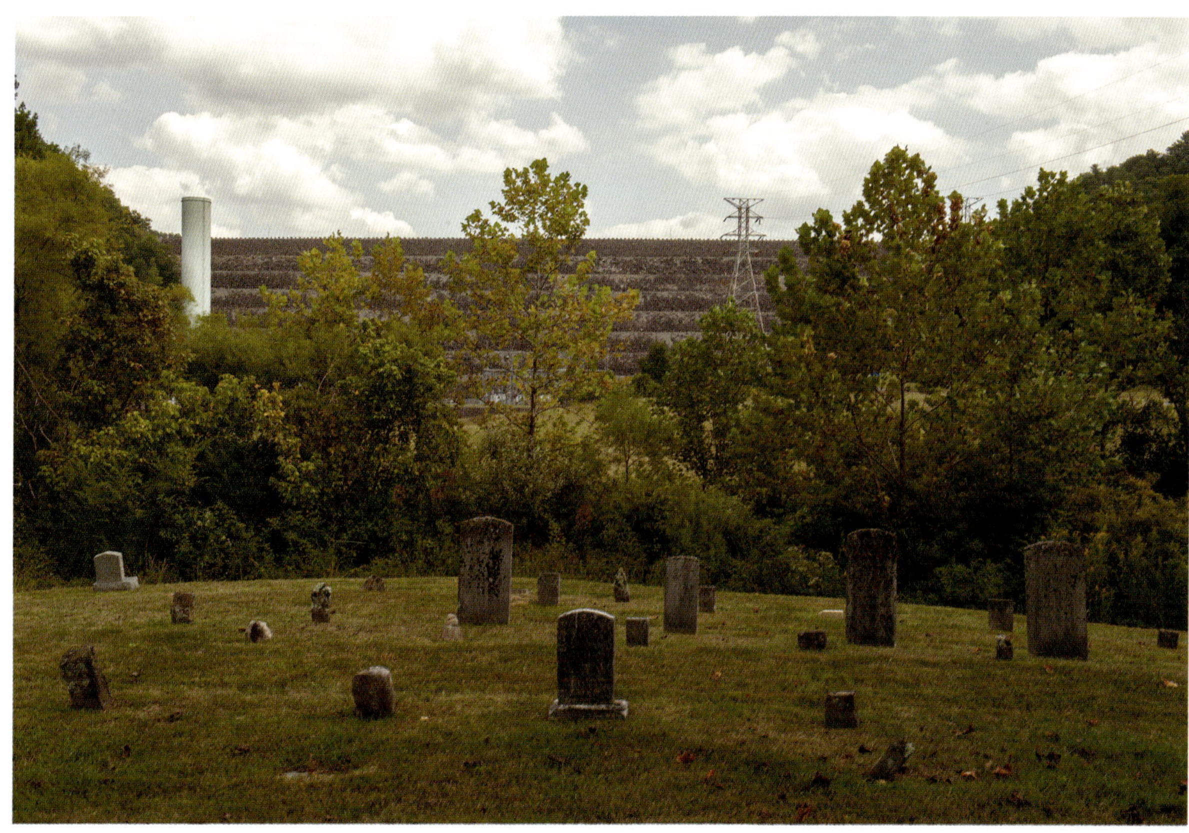

South Holston

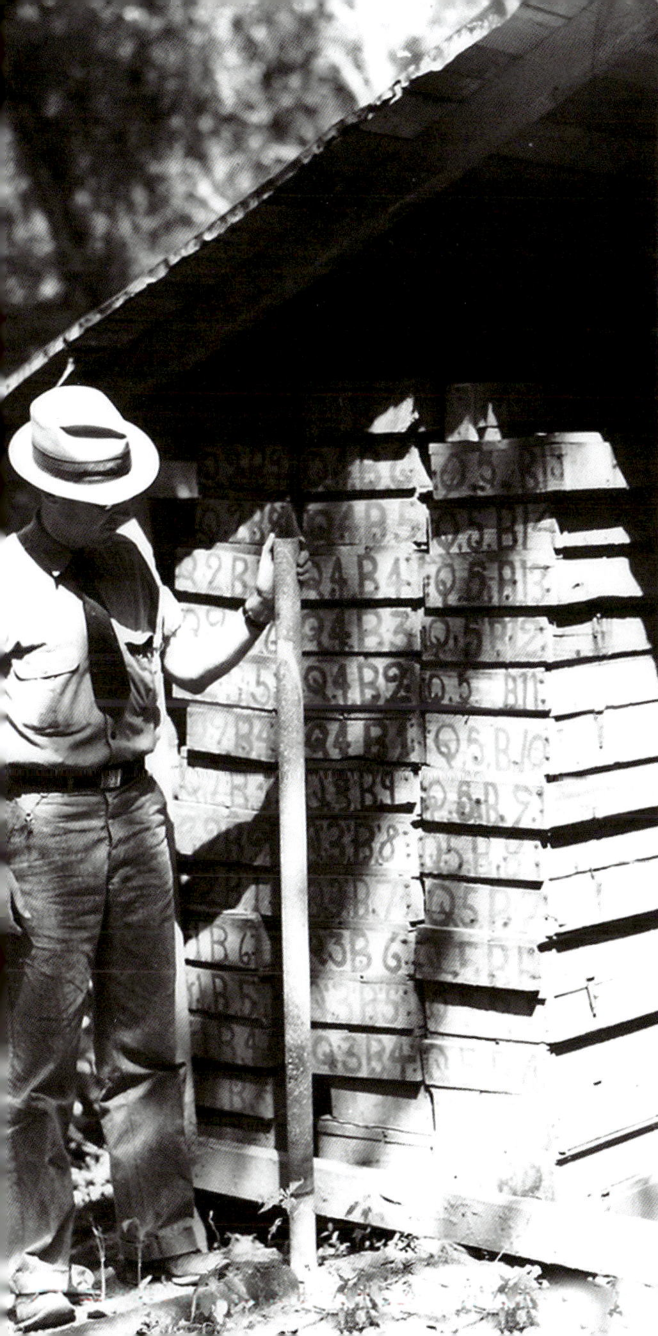

SECTION TWO
Southeastern Tributaries

Hiwassee Dam (1936) / Tennessee Valley Authority

Little Tennessee River

The hidden jewel of TVA's service area is the Little Tennessee River. Full of splendor and charm, the river boasts the tallest dam in the eastern United States along with one of TVA's most controversial hydroelectric projects. Beginning in north Georgia, the Little T (as it is referred to by locals) flows through western North Carolina, forming the southern boundary of Great Smoky Mountains National Park before meandering west to meet the Tennessee River south of Knoxville. While heavily impounded by a number of dams, all hydraulic flow control along the Little Tennessee River originates at Fontana Dam.

An icon of hydroelectricity as well as a significant tourist destination, Fontana Dam stands out as my favorite facility within TVA's multipurpose network of dams. Located in western North Carolina, the four hundred eighty-foot-high dam[1] is dwarfed by the loftier Smoky Mountains. Fontana's architecture borrows heavily from Norris Dam's modernist design, including sharp angles and inset overlooks. The construction announces itself with mid-century futurism.

Built during TVA's World War II construction initiative, it serves as one of the agency's most celebrated accomplishments due to its size and remote location. While not the most consequential TVA dam in terms of population displacement and grave relocation, the construction of Fontana flooded seven towns and a large portion of North Carolina Highway 288. The dam's construction remains a contentious subject within western North Carolina's social memory, with residents of Graham and Swain counties quick to express the pride in Fontana's ingenuity and unique engineering but dismay at how much land was taken by the federal government. Perhaps the most disheartened by the entire affair are the former residents of the North Shore.

The construction of Fontana Dam rendered several of the the newly-created lake's northern shore communities inaccessible. Acreage was acquired through eminent domain and residents were relocated; but, the cemeteries, which were located above the water line of Fontana Lake, remained. The newly acquired land, including the graveyards, was transferred to the Department of the Interior and later annexed into Great Smoky Mountains National Park.

In order to provide access to the family cemeteries, a plan to construct a new highway was agreed upon: the state of North Carolina would build a road from nearby Bryson City to the park boundary, and the National Park Service would construct a road along Fontana Lake's north shore, from the park's edge to the top of the dam. The state fulfilled its obligation in 1959; however, the park service only completed a portion of the road, shutting down construction in 1971 due to environmental concerns. Referred to by locals as the Road to Nowhere, the completed six miles of road ends at a 1,200-foot long tunnel that serves as the eastern entry point to the park's Lake Shore Trail.[2] A curious landmark for hikers, the tunnel is both a memorial to sacrificed land and a salient reminder of what many believe to be a broken promise.

That is not to say that the cemeteries are uncared for and unvisited—they are maintained by the National Park Service and tended by the remaining former inhabitants of the North Shore and their descendants. Annual decoration days are planned for each of the North Shore graveyards, and on select weekends from May through October, the National Park Service provides free boat transportation across the lake as the cemeteries are more easily accessible by boat.[3] I visited a few of these cemeteries during my fieldwork, and I can attest to the fact that many of the trails leading to them are not easy to hike. To the people of the Fontana region, what would normally be a simple act of remembrance has become an annual pilgrimage, one that honors the dead as well as the altered landscape they once called home. To local residents, Fontana is both a celebratory and a solemn dam. While this juxtaposition exists at every TVA location where population removal was necessary, it seems especially tragic and pronounced at Fontana.

I approached the area in and around Fontana's reservation with as much of a local's point of view as I could. I wanted to be mindful of the location as a wilderness destination—it is part of the Appalachian Trail—as well as a historical spot that symbolized both progress and sacrifice. *Fontana* (52) considers the dam as an obstacle, and the full height and width of the structure is withheld from the viewer, making it impossible to "reach" the mountains that rise beyond it. In an opposite approach, *Fontana, Visitor Center* (64) presents the dam from the controlled environment of the visitor center. The two photographs offer opposing views: one interprets the dam as an obstruction that cannot be traversed; while the other perceives Fontana as a monument to progress and nationalism, a structure to be visited, and an accomplishment to be viewed with awe and pride. From the opposite side of Fontana Lake, and in further contrast to TVA's narrative of Fontana Dam, *Fontana, Lakeview Drive* (66), contemplates the unfulfilled needs of a still grieving community. The pylons presented in this shot mark the end of the Road to Nowhere, and serve as reminders to the local community of what they can no longer easily access.

The Little Tennessee River is home to another contentious hydroelectric location: Tellico Dam. This dam's story is very different than that of Fontana's, including the fact that its existence was argued before the United States Supreme Court. The Tellico controversy can be best summarized as a standoff between ordinary citizens and special interest groups, with TVA, who had historically placed itself in the interests of everyday people, pulling a significant role-reversal.

The rationale behind many of TVA's dams was historically quite tangible: electricity, flood control, and navigation. With Tellico, however, the proposed benefits were speculative and hinged upon industrial development at the newly-created site. Construction began in 1967, but the dam's completion was stalled when local farmers, environmentalists, and Cherokee tribal leaders united in opposition and filed a lawsuit under the Endangered Species Act to protect a species of small fish called the Snail Darter. The case ultimately made its way to the United States Supreme Court which halted the closing of the dam's gates in 1978. In response, Congress passed a rider to an energy and water resources appropriations bill that exempted Tellico from the Endangered Species Act. The gates closed, and the reservoir began to fill in 1979. Ultimately, the consequences of Tellico's construction and lengthy controversy resulted in the damming of the last free-flowing section of the Little Tennessee River, a longer-lasting and deeper-seated indignation toward TVA by the very people the company claimed to serve, and the end of TVA's dam-building initiatives.[4]

The photographs found within this book are not meant to dwell on the past but instead to consider the past as a means of informing the present. That being said, I wrestled with how to photograph Tellico. Treating it like any other hydroelectric dam would be dishonest. I wanted to acknowledge its existence while being sensitive to the lasting effects and residual resentment it generated in the community. My solution was to capture images in the form of an elegy. The Tellico photograph is a memorial not only to the people that were dispossessed and to the lost wildness and wildlife of the Little Tennessee River but also to the Tellico situation as a whole.

Tellico, Reservoir (66) contemplates the legal, ecological, corporate, and social debacle that the construction of this seemingly insignificant dam became. To people unaware of Tellico's history, this photograph might be seen as a fairly pedestrian image of a reservoir, but its unassumingness was necessary and purposeful: Tellico Dam and Fontana's Road to Nowhere define these landscapes differently to their local populations and continue to inflict grief and sadness on those affected.

Hiwassee River Basin

The Hiwassee River basin is the most southern of the Tennessee River's main tributary systems and contains a variety of moderately-sized flood control dams. Overall, there are eight TVA-operated hydroelectric facilities within the region, including the beautifully designed Hiwassee Dam. Wedged into a steep mountain valley, the dam towers over its powerhouse. It stands over three hundred feet tall[5] and is topped with an impressive gantry crane. While TVA's uniquely designed gantry cranes are found at many of the agency's larger dams, the one atop Hiwassee fascinated me the most because of its accessibility. As seen in the photograph, *Hiwassee* (59), the crane is fully accessible to pedestrian and vehicular traffic, and appears out of place among the surrounding mountains. Another unique mix of industry and nature can be seen further downstream at Apalachia Dam.

A World War II hydropower project, Apalachia's reservation is adjacent to Nantahala National Forest and possesses minimal recreational activities other than backwoods camping. Its defining feature is a pipeline that juts out from the dam's concrete face and disappears into the opposing mountainside, delivering water to the Apalachia powerhouse 8.3 miles away. *Apalachia, Reservation* (53) was made on my final visit to the dam in August of 2015 and captures water streaming through what appears to be a bullet hole in the pipe. Inexplicably, a family was spending the afternoon on the sparse reservation, wading in the water downstream and relaxing along the river's banks. Three children treated the punctured pipe like a waterfall and danced under its steady stream without any concern that they were playing under a massive industrial pipeline. What I saw left me curious, as there is no TVA-designated recreational space on Apalachia's reservation, as well as no place to park a car other than along the access road. I chose not to photograph the family, as they seemed uncomfortable by my presence, and I did not want to intrude. I did speak briefly with the children's grandfather, who was clearly a hydroelectric tourist. I asked him why he had brought his family to Apalachia Dam when there were other, larger dams close by.

"Why not?" he said with a smile. "It's here isn't it?"

His answer reveals much about the people of the Tennessee Valley and their unwavering sensibility to take agency within the spaces that they technically own. It is common to see people fish, swim, and pass the time wherever they please in this region. The dams are treated with respect and utilized as they see fit, whether that utilization is in accordance with TVA's designation or not. My interaction with the family at Apalachia was, in fact, similar to an experience I had a year earlier when I photographed *Chatuge, Reservoir* (55).

The setting was similar: A family was playing and swimming at a location with no regard for its intended designation. They were at a simple embankment and seemed entirely accepting of the fact that there was no beach sand, no stairs, and no floating boundaries. There was only wet, rocky, red clay. What was puzzling about the situation was that a

TVA-designated beach could be seen from where the family stood, yet they had driven their car out of the nearby parking lot and down the lakeshore to this location. I considered approaching the woman to ask why she and her children were not at the nearby public beach, but she regarded me with suspicion. Instead, I hung back and made the photograph.

While it would be a stretch to suggest that Chatuge Dam is a tourist destination, it does offer substantial recreational opportunities. However, it pales in comparison to the Ocoee River. Dammed in three locations, the Ocoee River is home to the Ocoee Whitewater Center. Managed by the U.S. Forest Service, this stretch of river hosted the 1996 Summer Olympic Games whitewater slalom competition and now primarily serves the commercial rafting industry.

Beginning directly downstream of Ocoee Dam No. 3, rafters travel through the former Olympic course and float downstream to Ocoee Dam No. 2. As seen in the photograph *Ocoee #2* (51), the structure acts as a waypoint for the rafting companies as rafts must be removed from the river at the top of the dam and walked down to its base for the next put-in. This setting is a remarkable arrangement of altered landscape, hydroelectric engineering, and river recreation, and the Ocoee River is one of those rare locations where the transformative power of water can be witnessed regularly. TVA quite literally turns the river "on" during rafting days, and what is simply a shallow, rocky stream on non-rafting days quickly becomes a roiling recreational river. TVA enables the tourism of this region through sustainable economic support, not by providing cheap and abundant electricity, but instead by releasing water from a dam. In this way, dangerous waters bring joy and excitement to people who gather together to float down them.

NOTES

1. "Fontana," Tennessee Valley Authority, https://www .tva.com/Energy/Our-Power-System/Hydroelectric /Fontana-Reservoir.
2. Lance Holland, *Fontana: A Pocket History of Appalachia* (Robbinsville: Appalachia History Series, 2001), 187–92.
3. Alan Jabbour and Karen Singer Jabbour, *Decoration Day in the Mountains* (Chapel Hill: University of North Carolina Press, 2010), 112–15.
4. Hargrove, *Prisoners of Myth*, 171–77.
5. "Hiwassee," *Tennessee Valley Authority*, https://www .tva.com/Energy/Our-Power-System/Hydroelectric /Hiwassee-Reservoir.

Ocoee #2

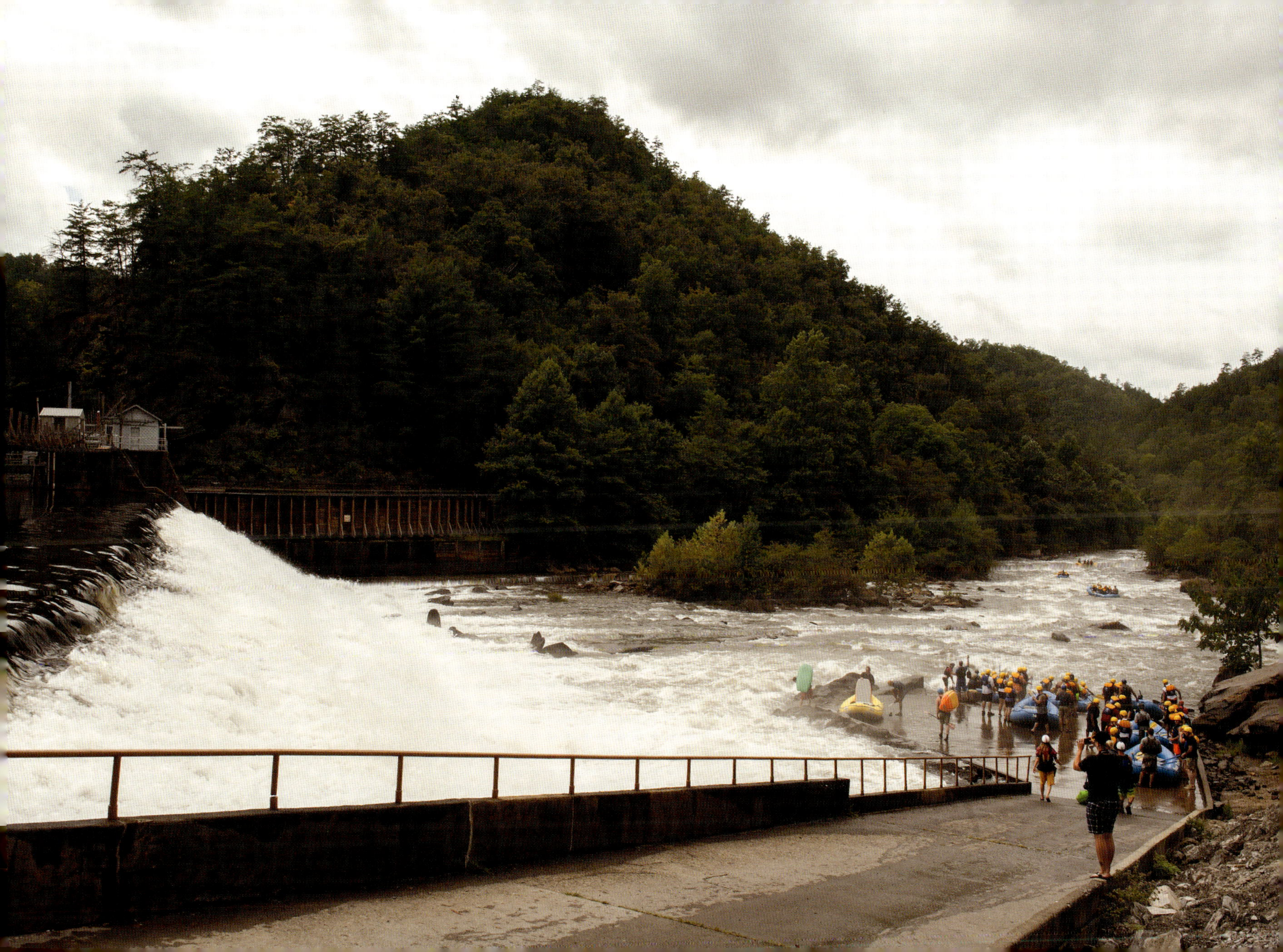

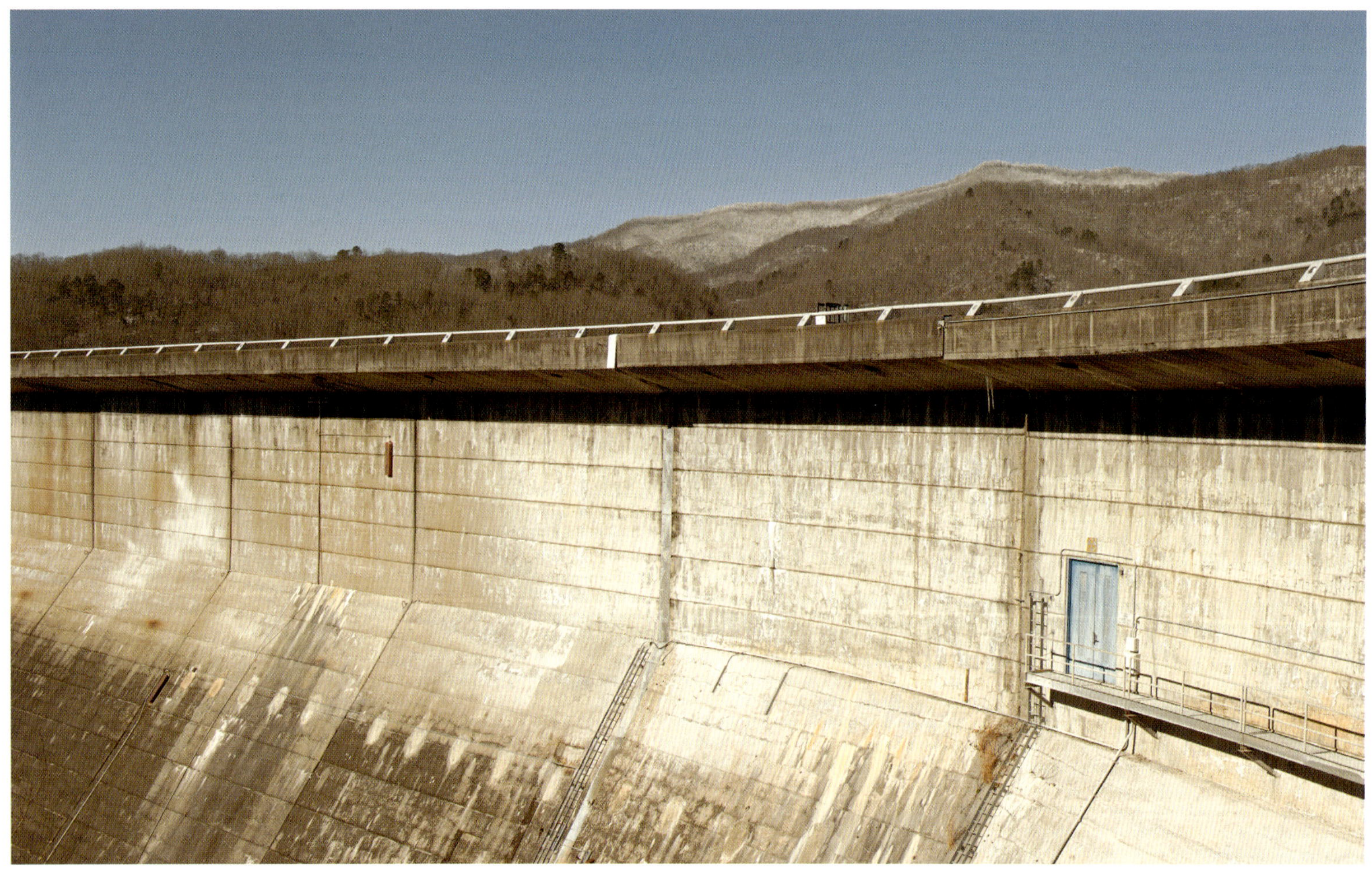

Fontana

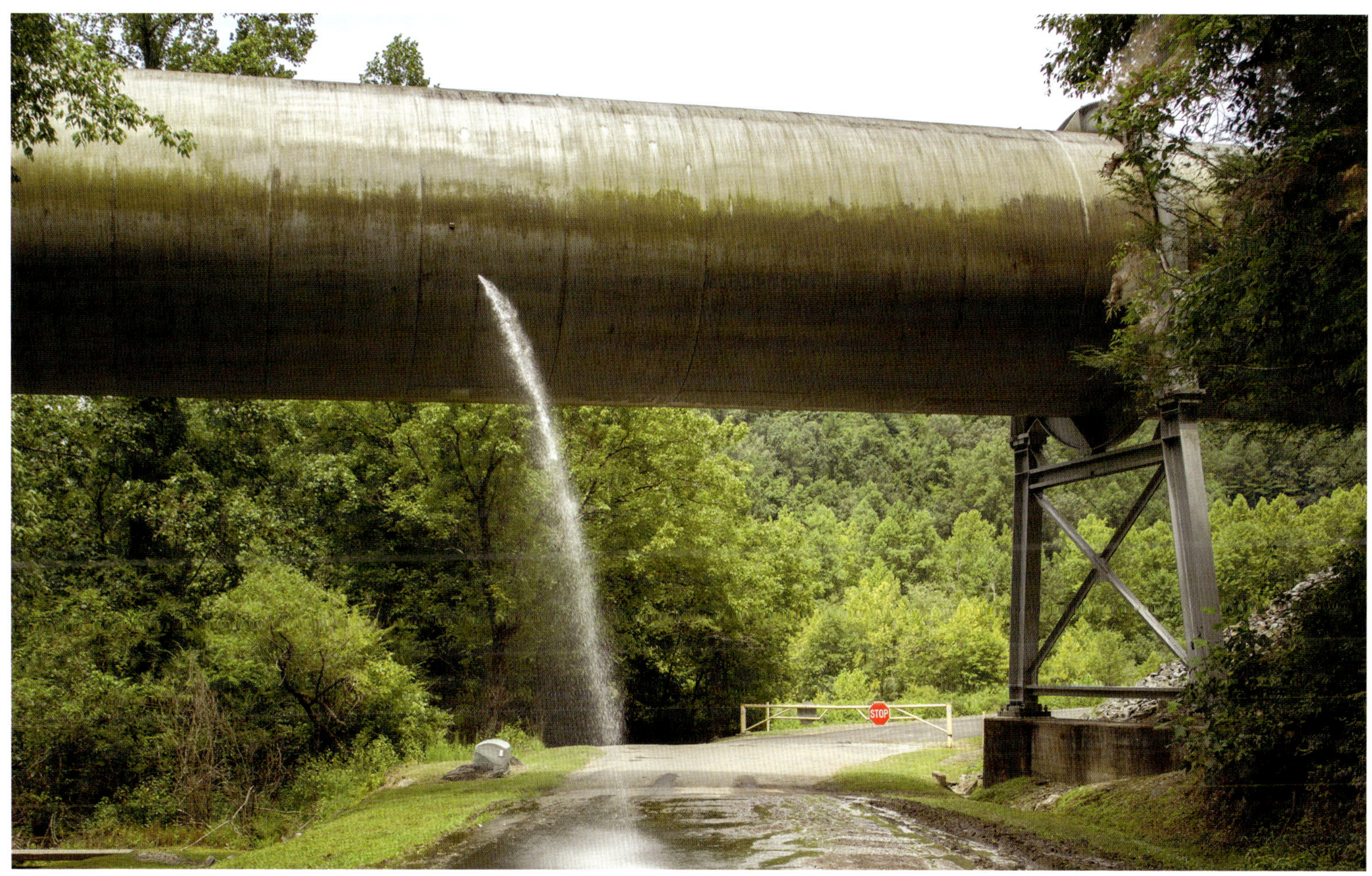

Apalachia, Reservation

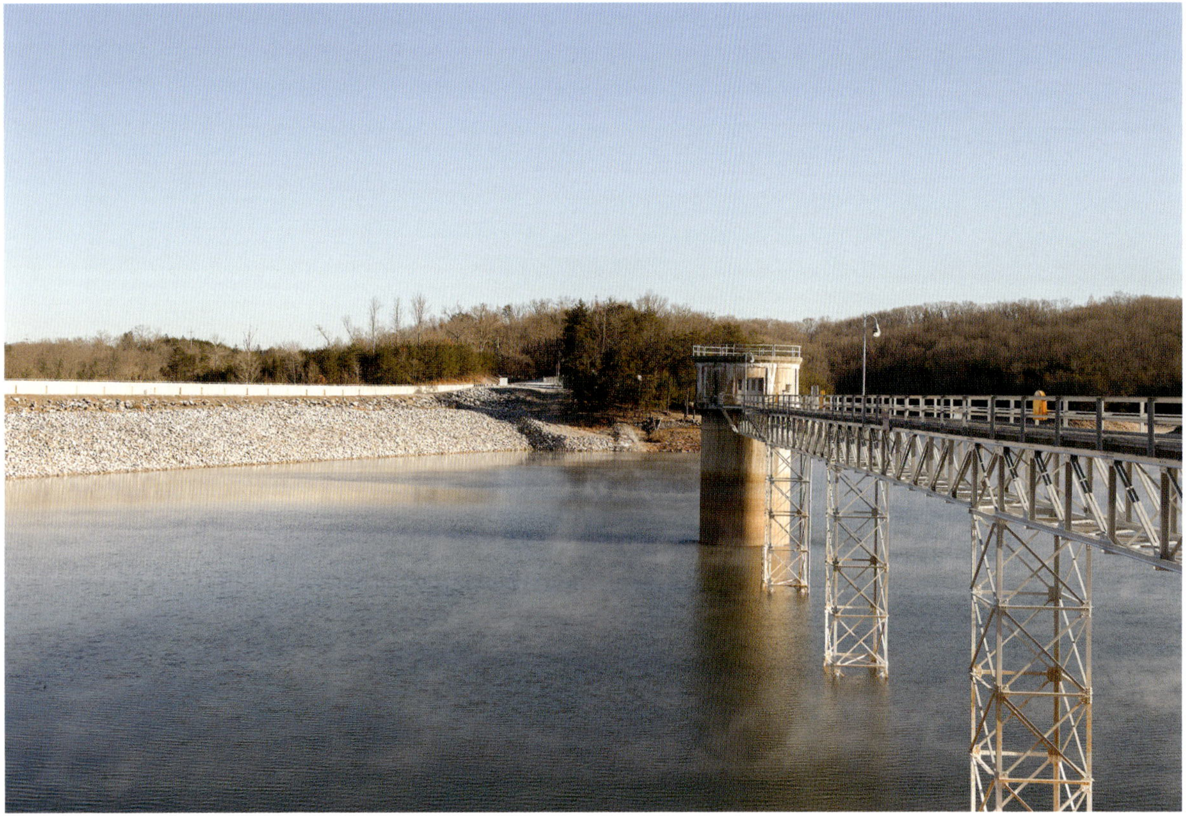

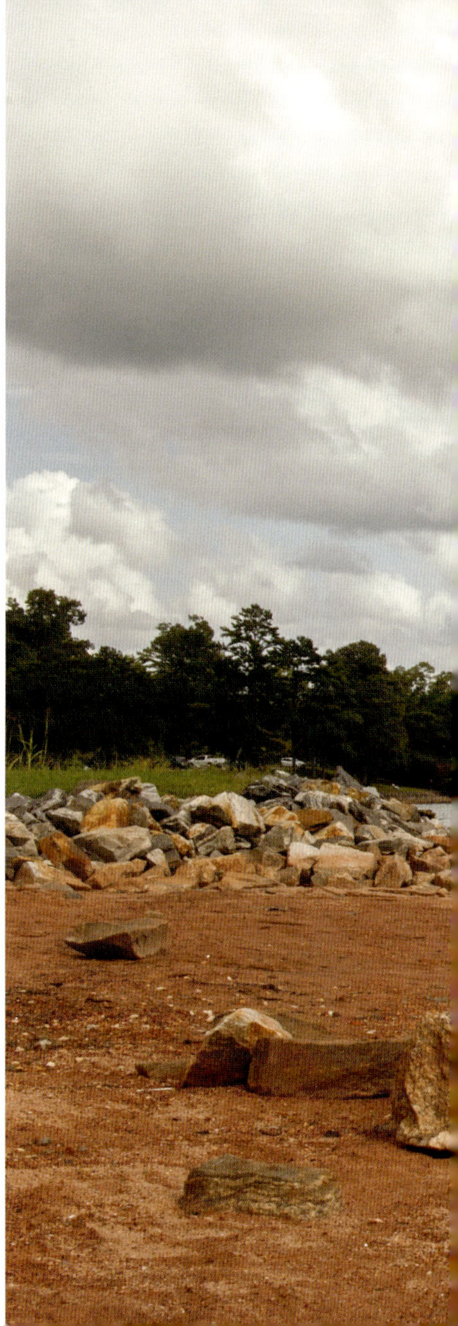

Blue Ridge

(facing) Chatuge, Reservoir

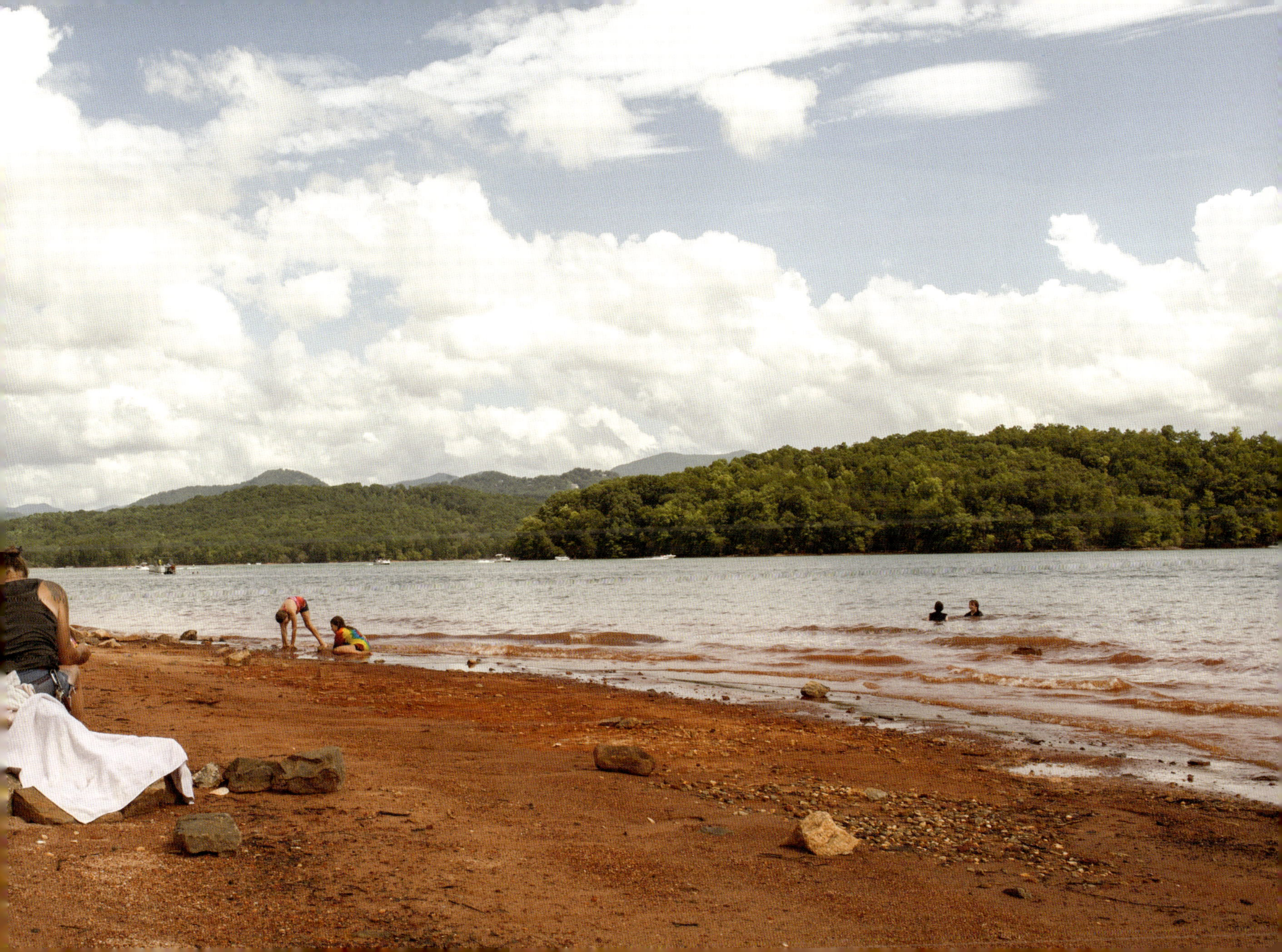

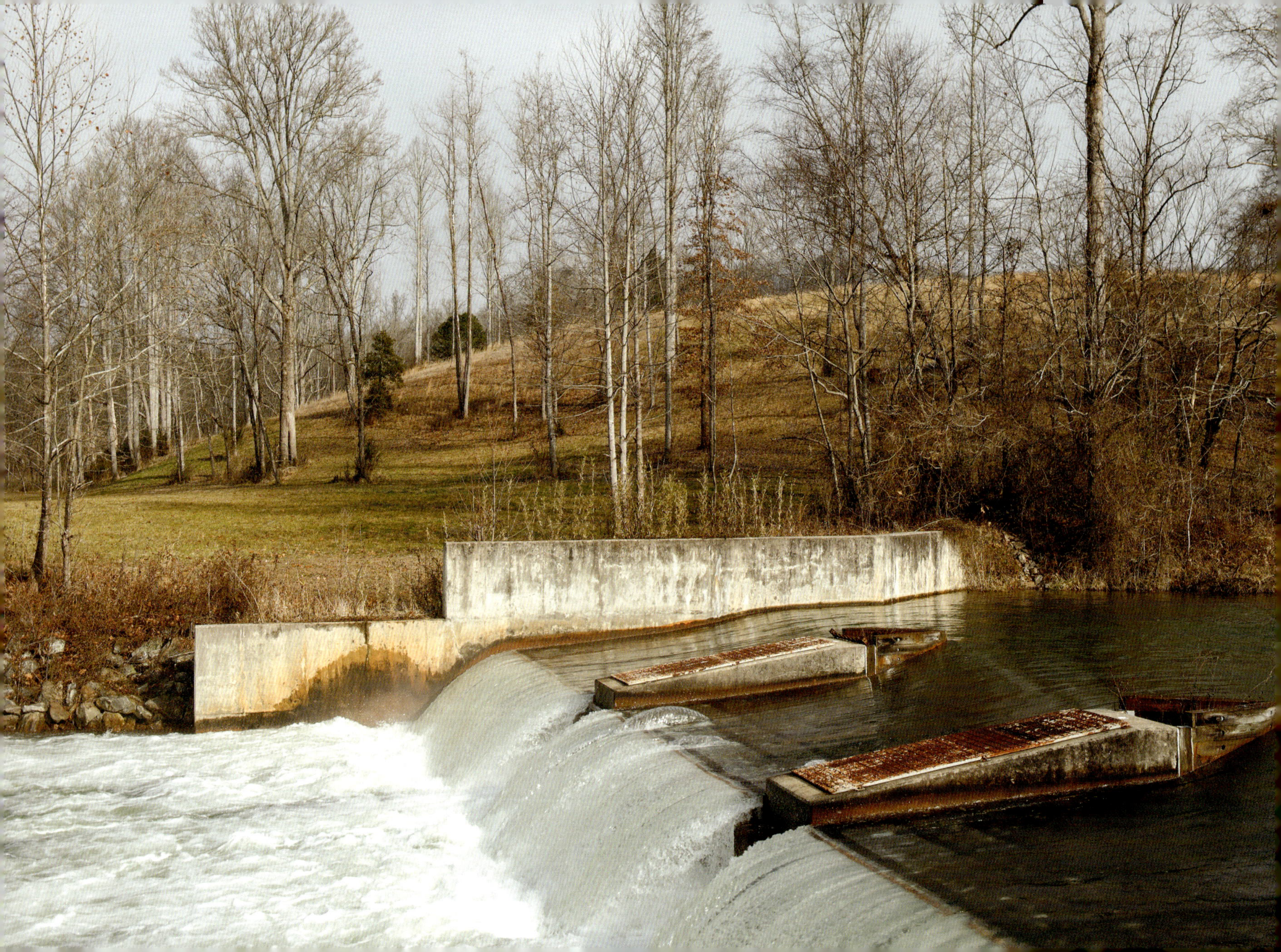

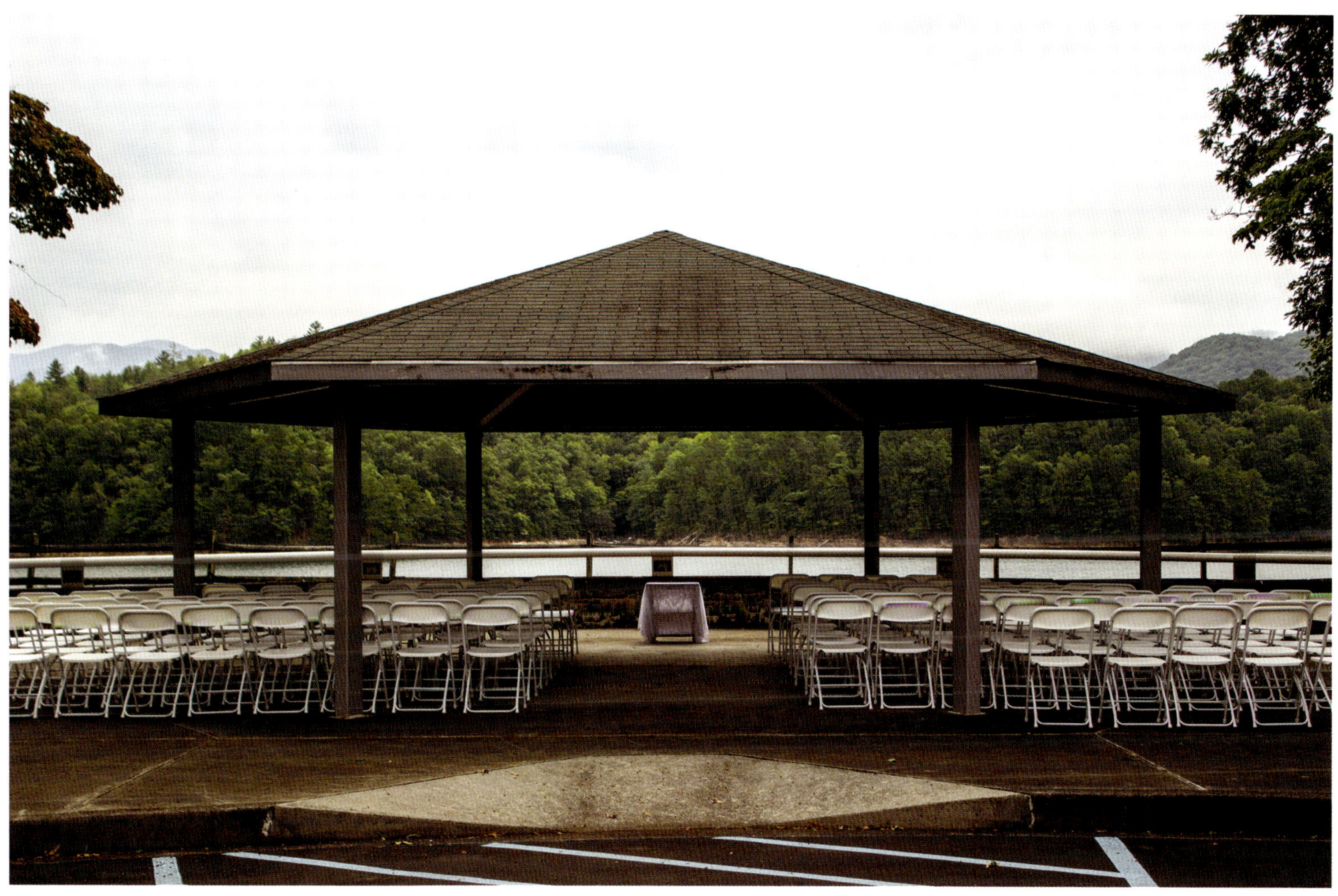

Fontana, Reservoir Overlook

(facing) Chatuge, Infuser Weir

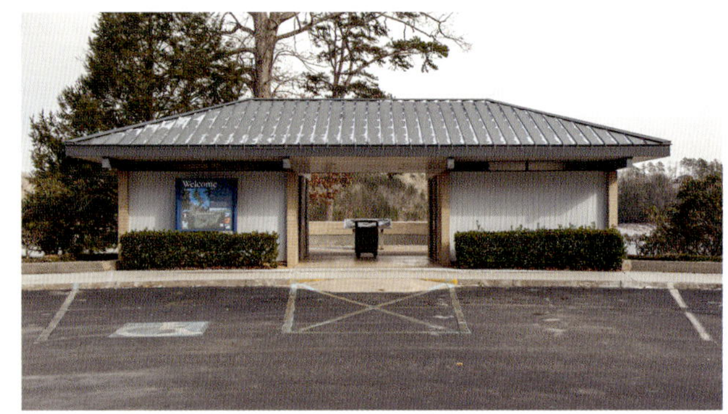

Hiwassee, Reservoir Overlook

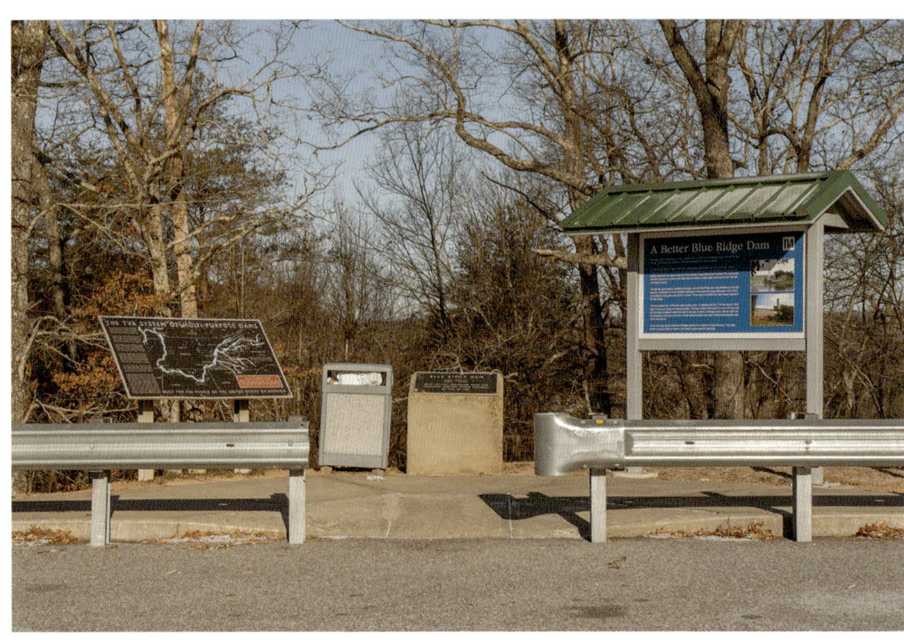

Blue Ridge, Picnic Area

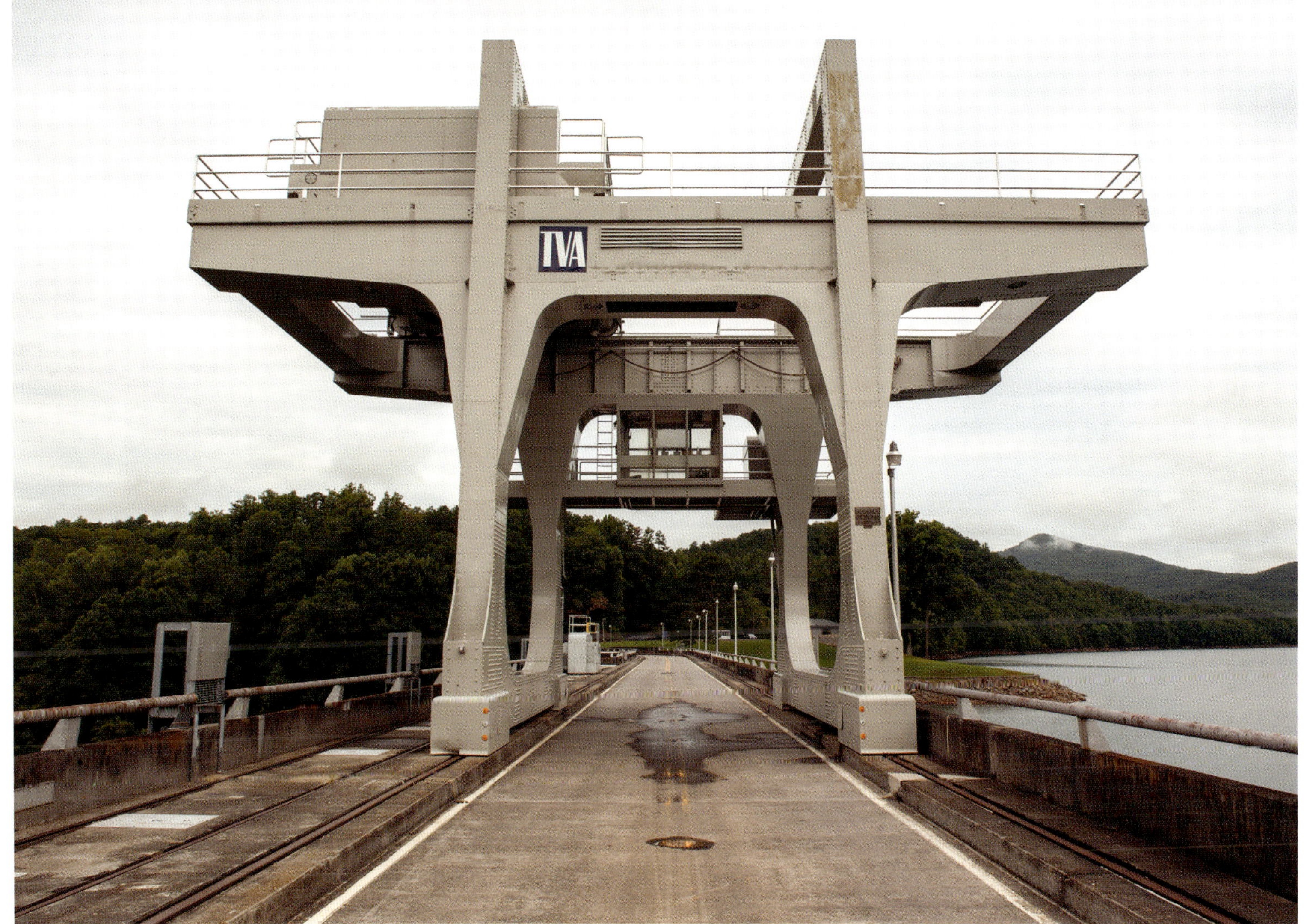

Hiwassee

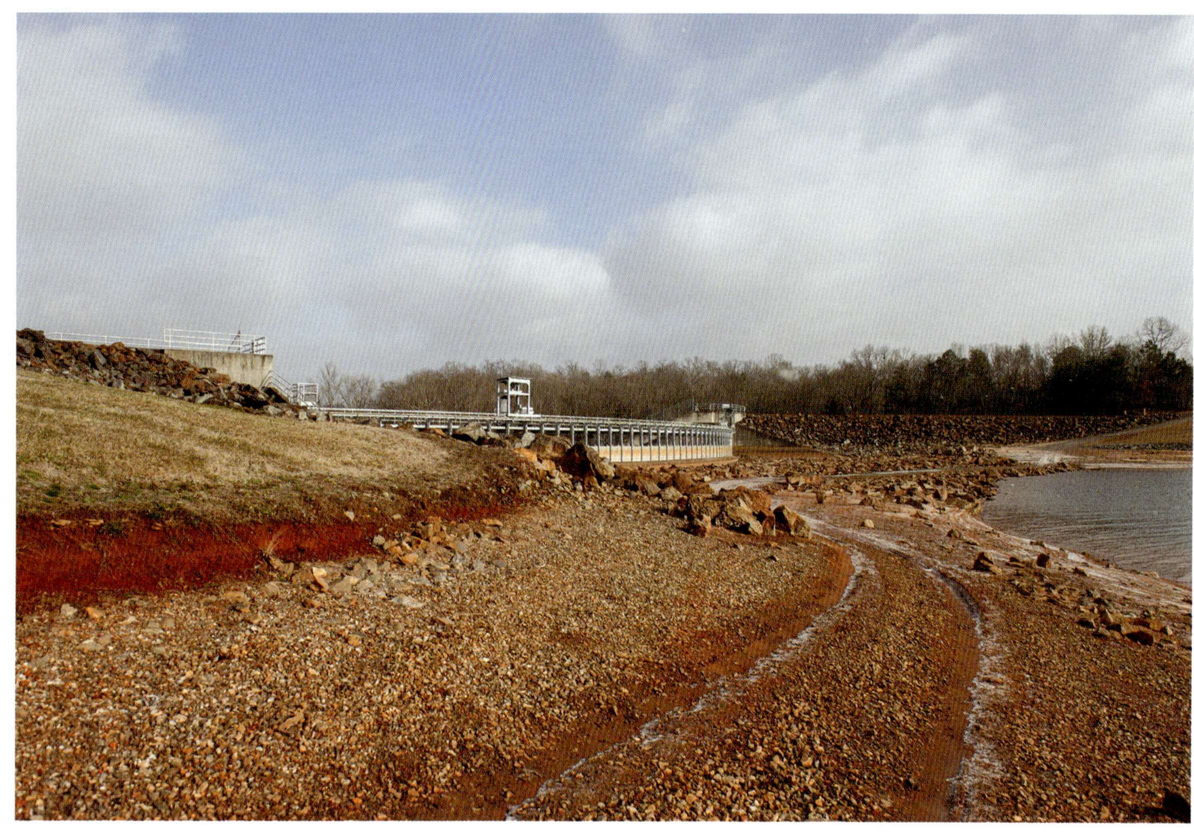

Chatuge

(facing) Apalachia

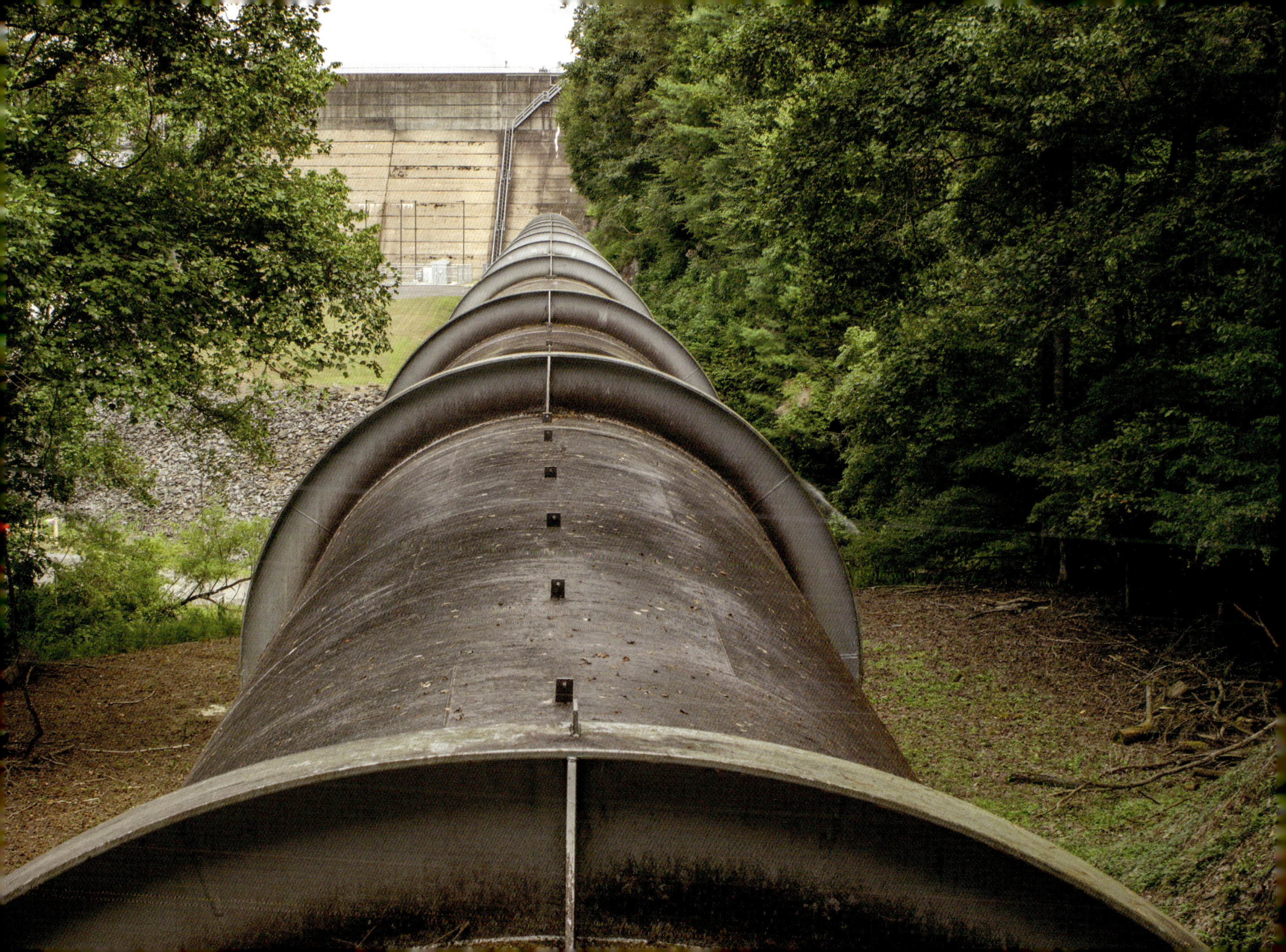

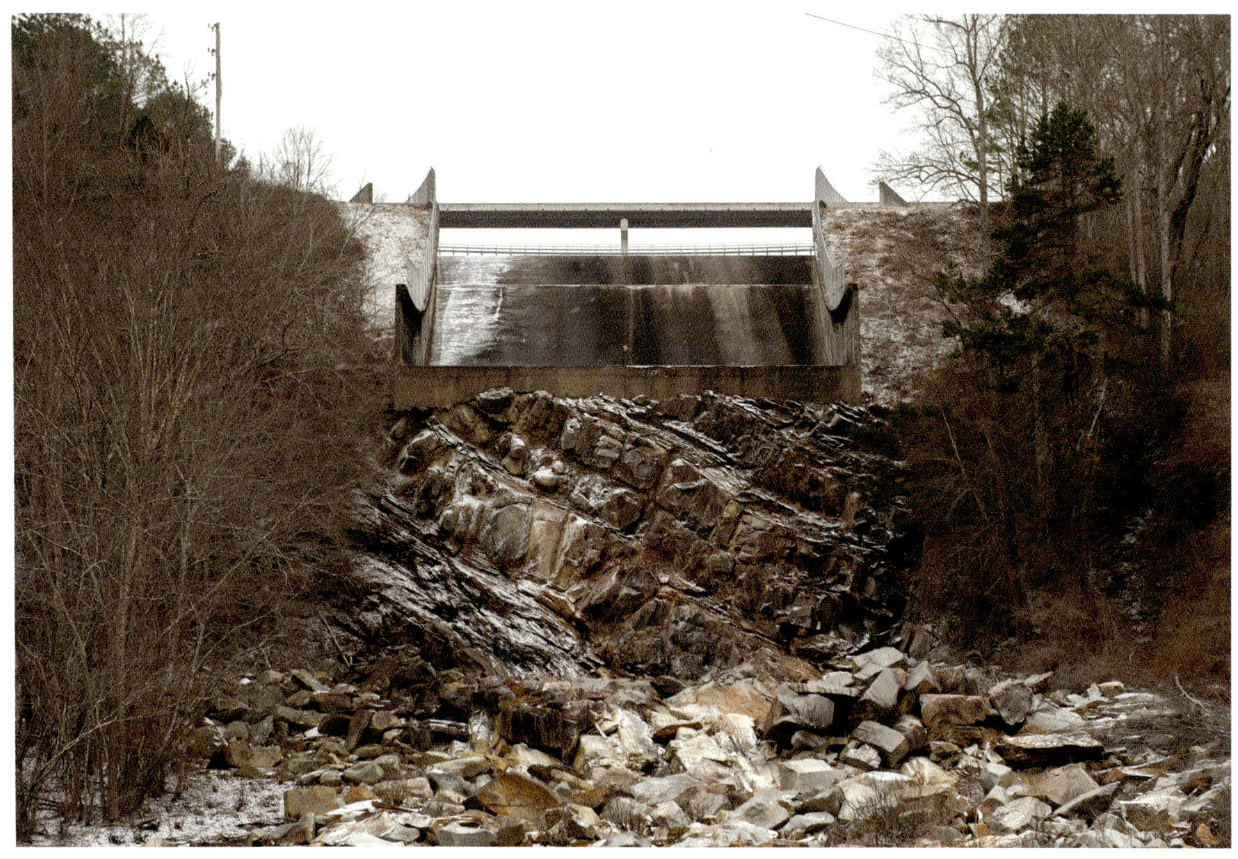

Nottely

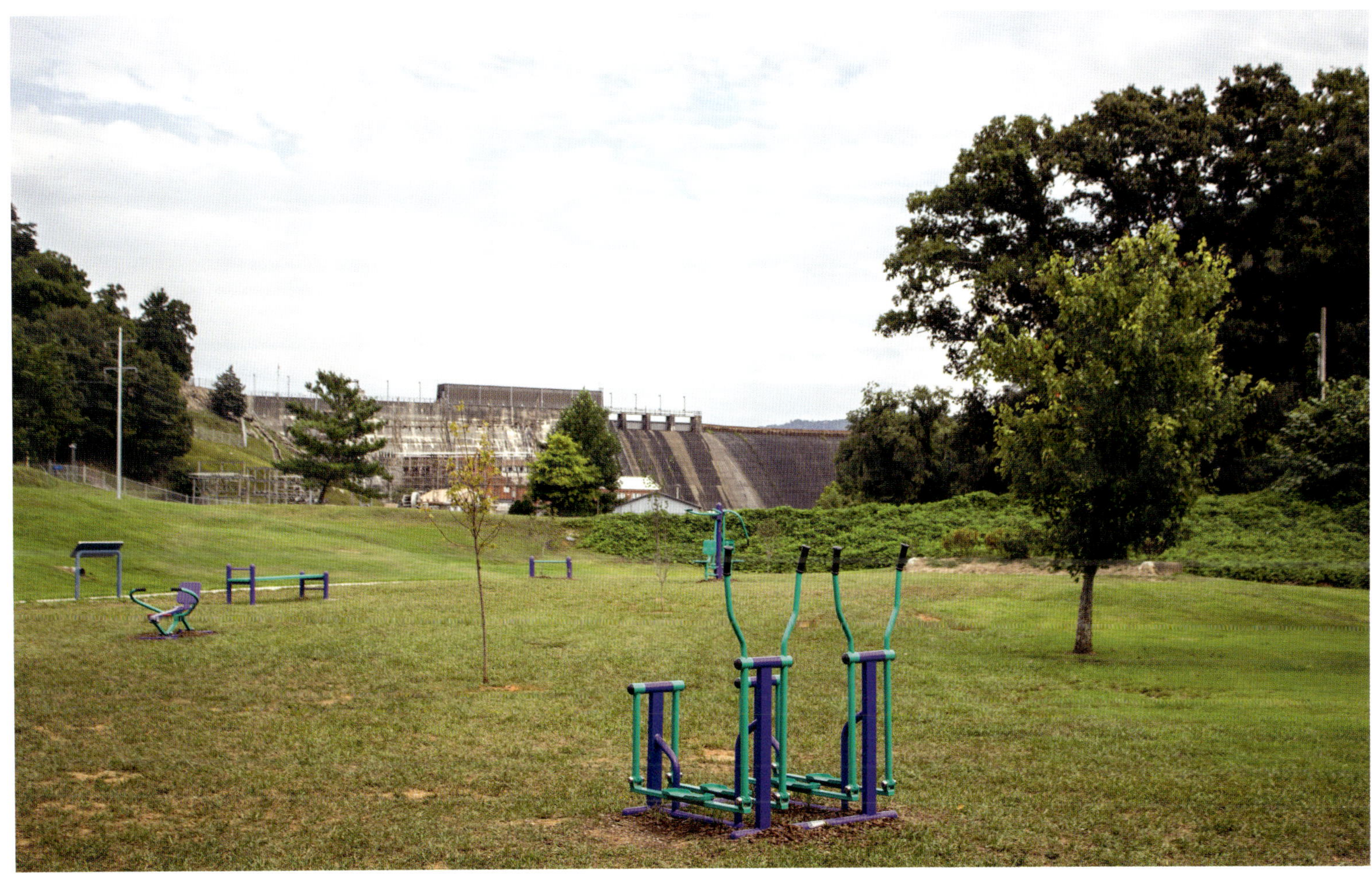

Ocoee #1

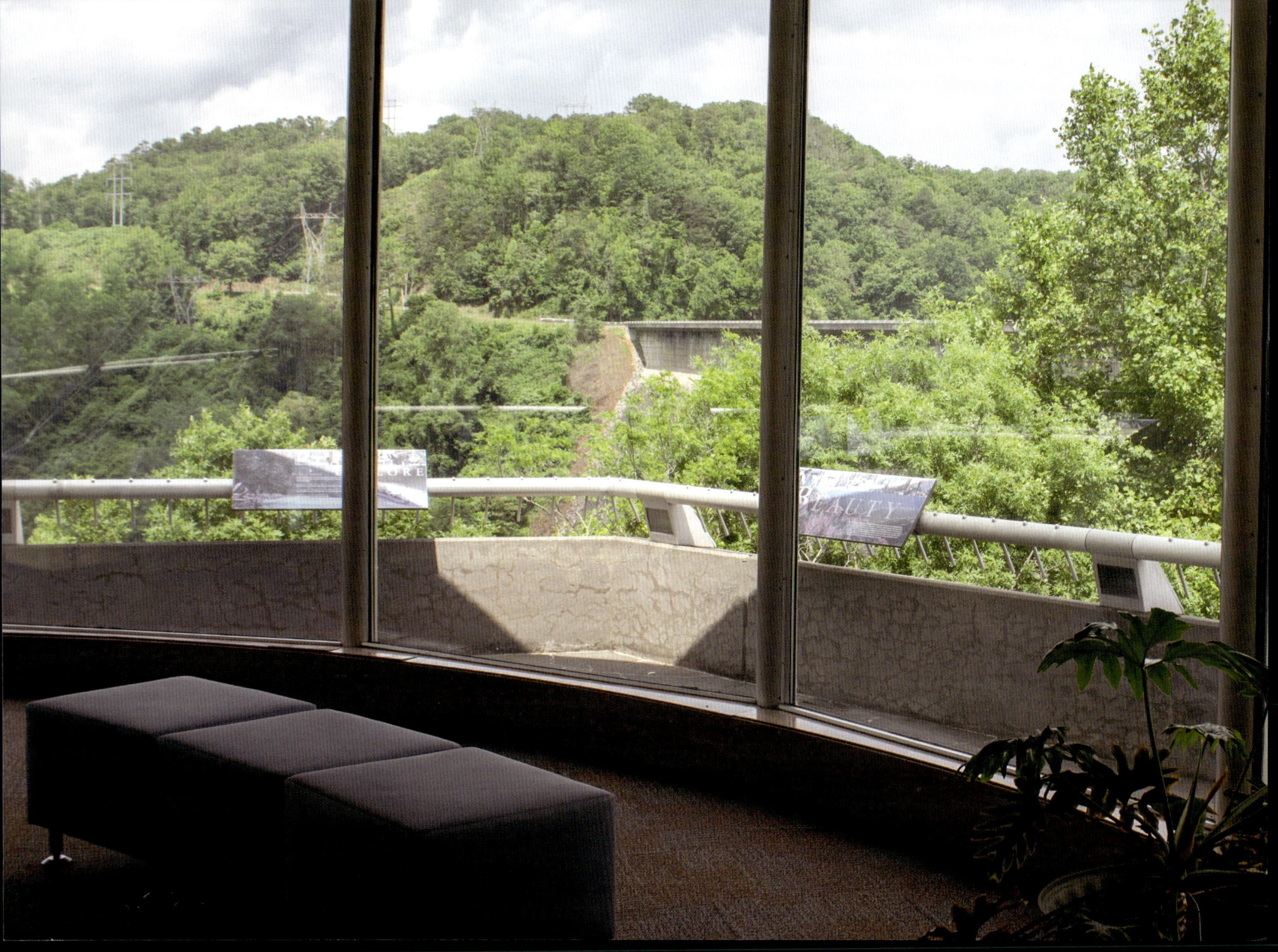

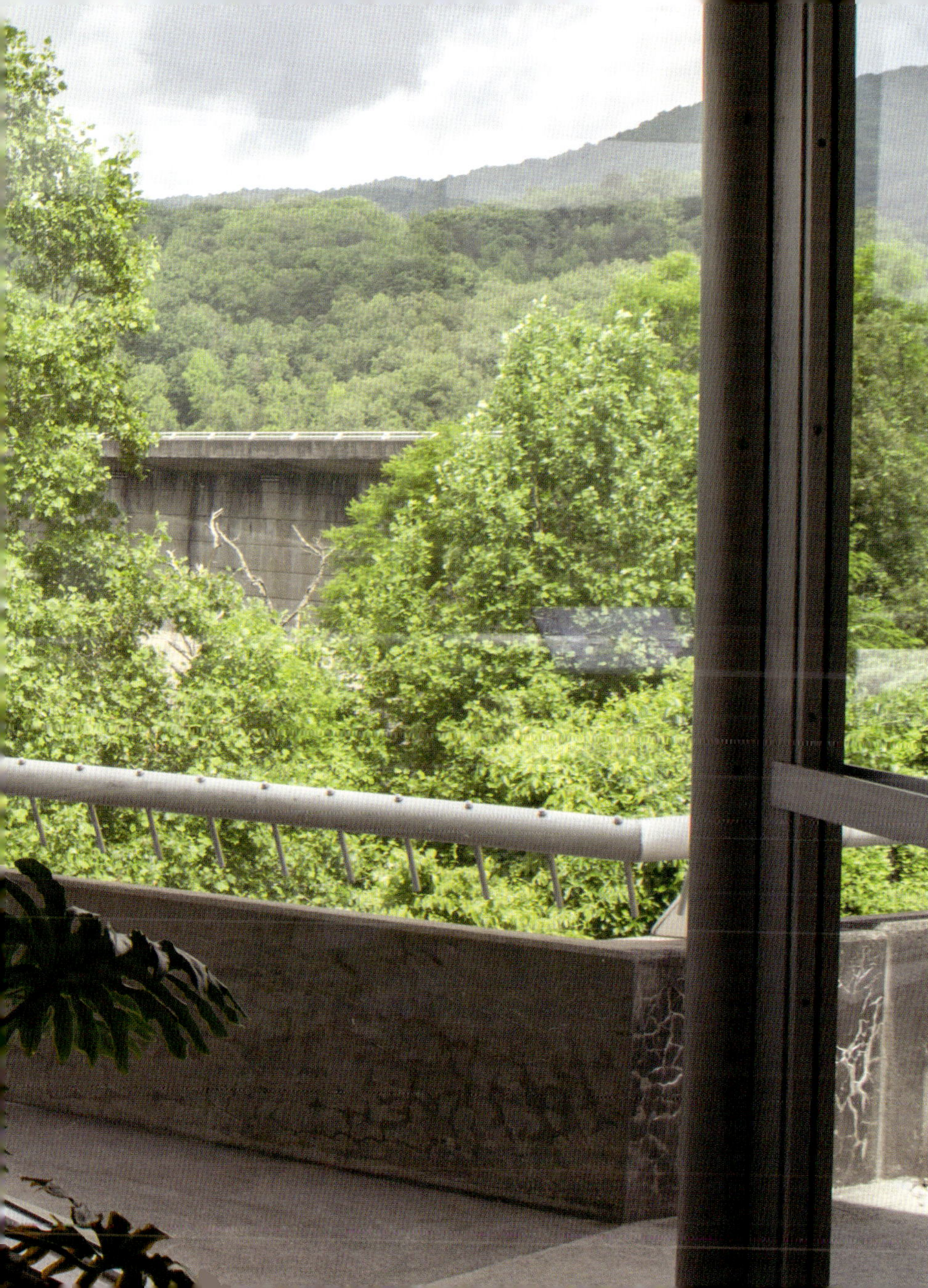

Fontana, Visitor Center

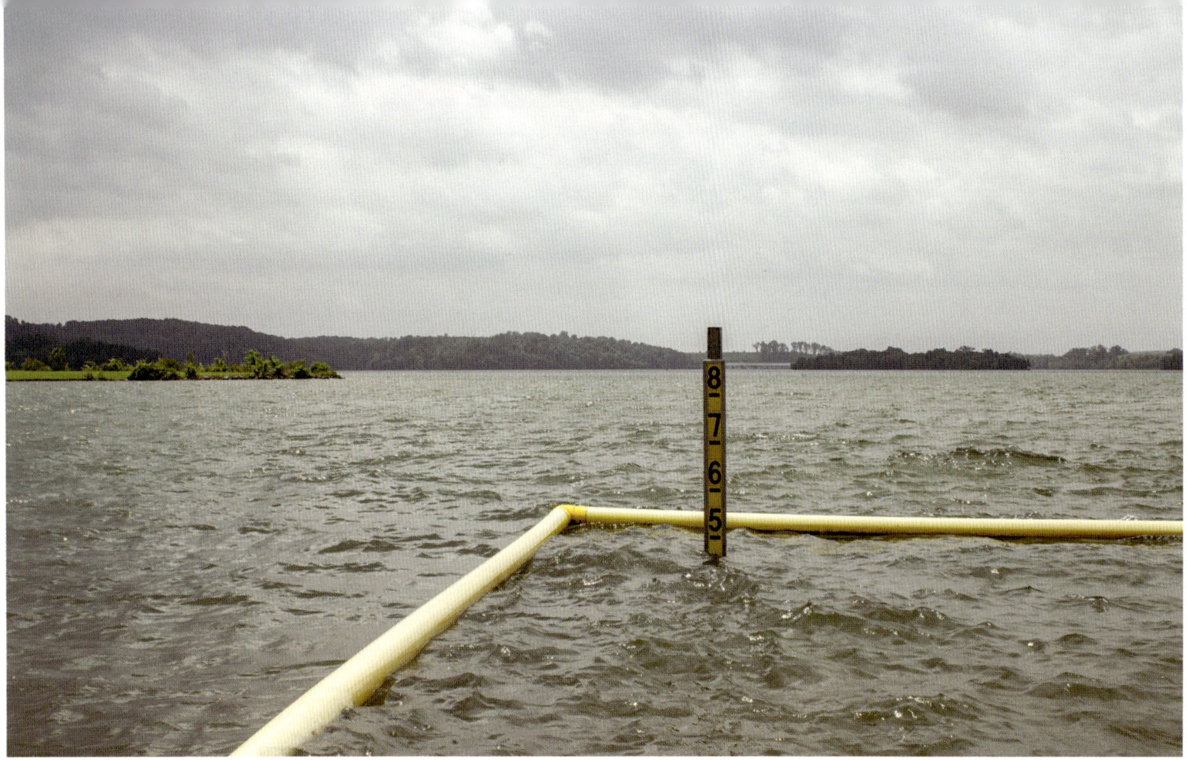

Tellico, Reservoir

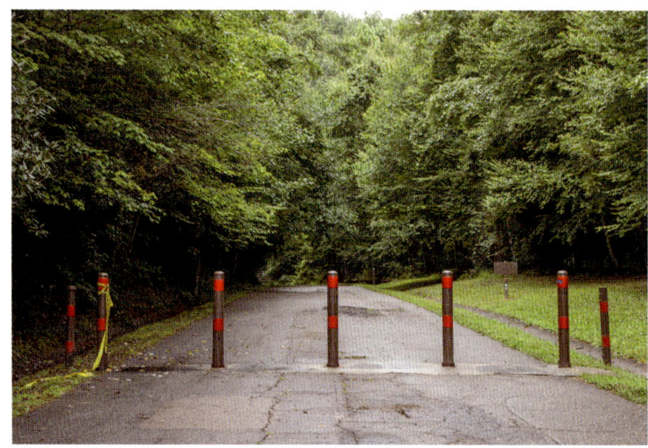

Fontana, Lakeview Drive

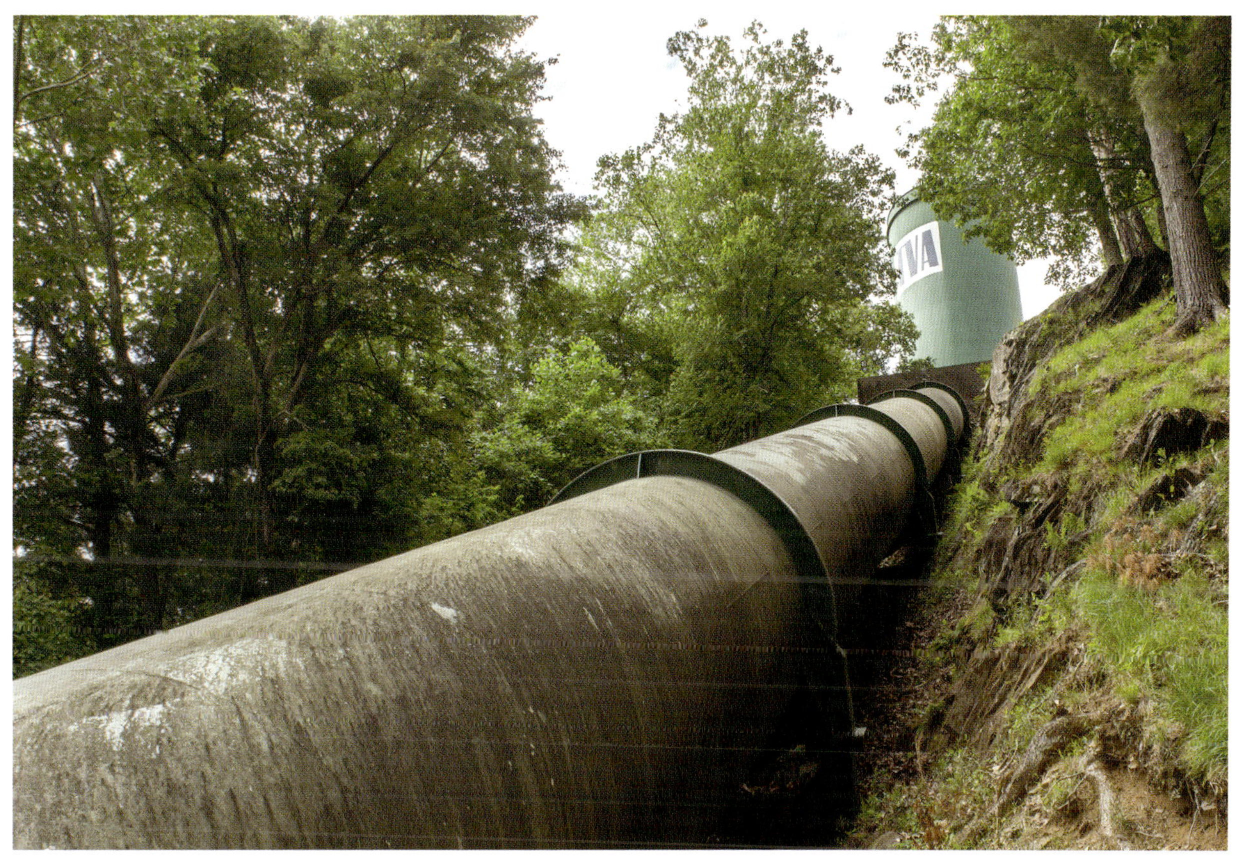

Ocoee #3, Powerhouse

Steamboat (1936) / Tennessee Valley Authority

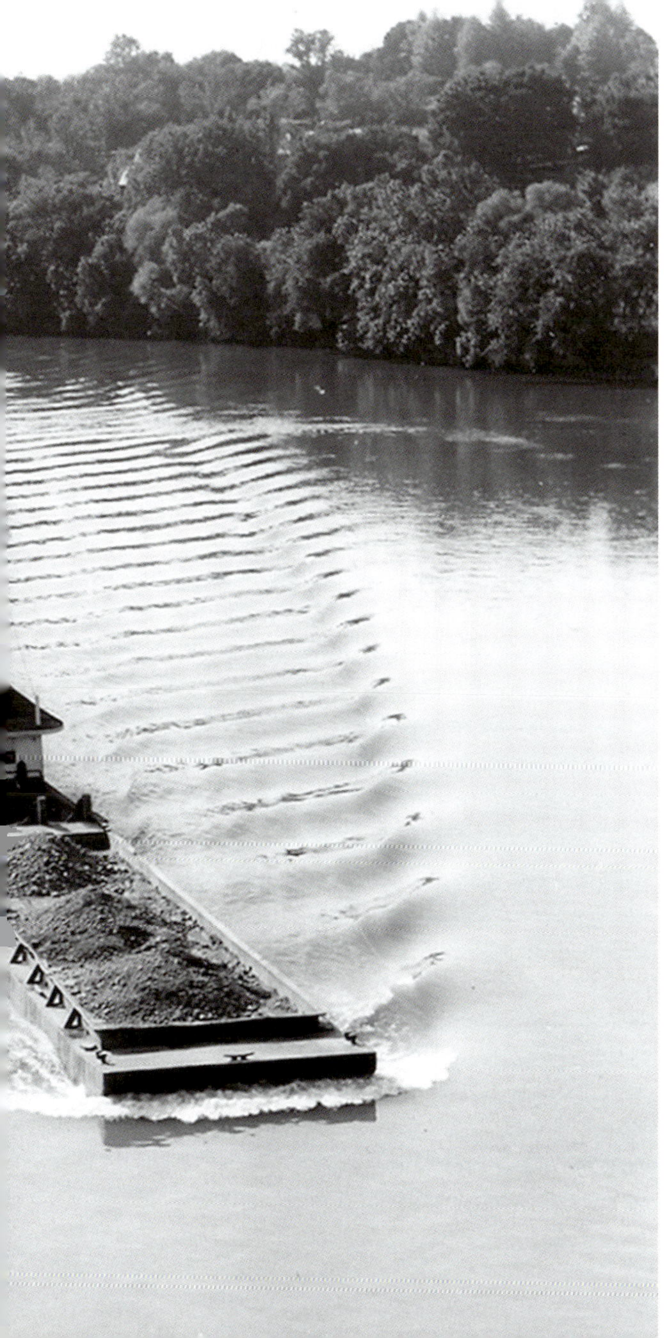

SECTION THREE

Central and Western Reservations

TVA manages a range of hydroelectric facilities in central Tennessee as well as a variety of flood control dams in the western section of its service area, which includes northwestern Alabama. Many of these flood control dams are small and remote, but their reservations are attended to with the same attention to detail and thoughtfulness for public recreation seen at the majority of TVA's larger facilities. My experience at these locations was consistently peaceful and pleasant, and despite their smaller footprint, they were comprised of well-maintained boat ramps and occasional picnic areas. Architecturally, everything was simple and practical. Constructed during the final decades of TVA's dam building initiatives, these locations deliver the same sentiment found within TVA's post-war facilities but lack the designed emotional punch delivered at locations such as Hiwassee and Norris. The exception is Great Falls Dam, which was completed in 1916 and acquired by TVA in 1939. [1]

Found in central Tennessee, Great Falls Dam is the only TVA-managed dam explicitly located outside of the Tennessee River watershed. Approximately seventy-five miles southeast of Nashville, its reservation shares land with Rock Island State Park and impounds the Caney Fork River. The location is renowned for its variety of recreational opportunities and beautiful landscape, including the thirty-foot-high falls that give the facility its name. The dam does not possess a true public overlook or any official visitor access other than a small roadside pull-off. Instead, the focus is on the the area's surrounding amenities and features which include the waterfalls, a nineteenth century textile mill, and the Caney Fork River Gorge. This minimized public display inspired me to explicitly photograph the dam itself and withhold its topographic context.

Central Tennessee is also home to both Tims Ford and Normandy dams. Tims Ford was constructed from 1966–1970 and impounds the Elk River. Built primarily for flood control, it stands one hundred and seventy-five feet tall and possesses one generating unit. [2] The dam's powerhouse can be seen in the photograph *Tims Ford* (page 76), with its black windows oddly placed within the surrounding walls of rock, and its spillway cutting through the landscape to the right. Normandy Dam was completed in 1976 and is similar to Tellico Dam in design.[3] It has no hydroelectric generators and exists solely as a flood control facility. Its reservation contains a popular boat launch and a large parking lot downstream of the dam. Overall, Normandy's public access and functional design is similar to the cluster of TVA dams found in northwest Alabama.

The four dams along Bear Creek and its tributaries lie within close proximity to each other and were also built primarily for water retention and flood control. None of the dams include hydroelectric components, but they do possess standard public facilities such as parking lots and boat ramps. The photograph *Little Bear Creek, Fishing Pier* (page 80–81) presents one of the more thoughtfully designed and well-maintained facilities in the area, while *Bear Creek, Recreation Area* (page 78) documents a location that is little used and in need of refurbishment. Overall, these four facilities display the breadth of TVA's hydroelectric program in this area, as well as the agency's thorough commitment toward design, public access, and outdoor recreation, even at the smallest and most remote facilities.

NOTES

1. "Great Falls," *Tennessee Valley Authority*, https://www.tva.gov/Energy/Our-Power-System/Hydroelectric/Great-Falls-Reservoir.
2. "Tims Ford," *Tennessee Valley Authority*, https://www.tva.gov/Energy/Our-Power-System/Hydroelectric/Tims-Ford-Reservoir.
3. "Normandy," *Tennessee Valley Authority*, https://www.tva.gov/Energy/Our-Power-System/Hydroelectric/Normandy-Reservoir.

(facing) Great Falls

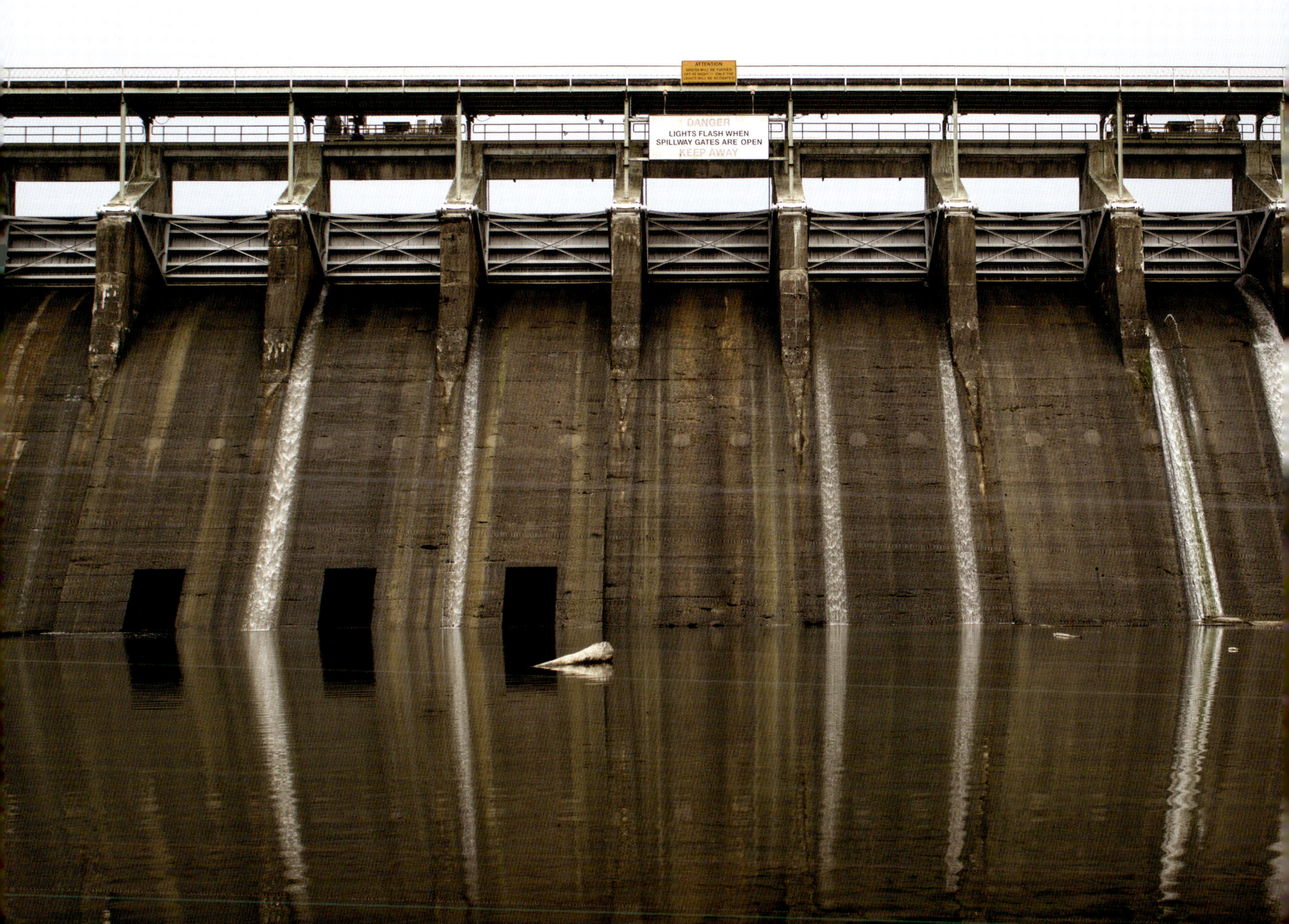

ATTENTION
SIRENS WILL BE TURNED
OFF AT NIGHT ONLY THE
LIGHTS WILL BE ACTIVATED
DANGER
LIGHTS FLASH WHEN
SPILLWAY GATES ARE OPEN
KEEP AWAY

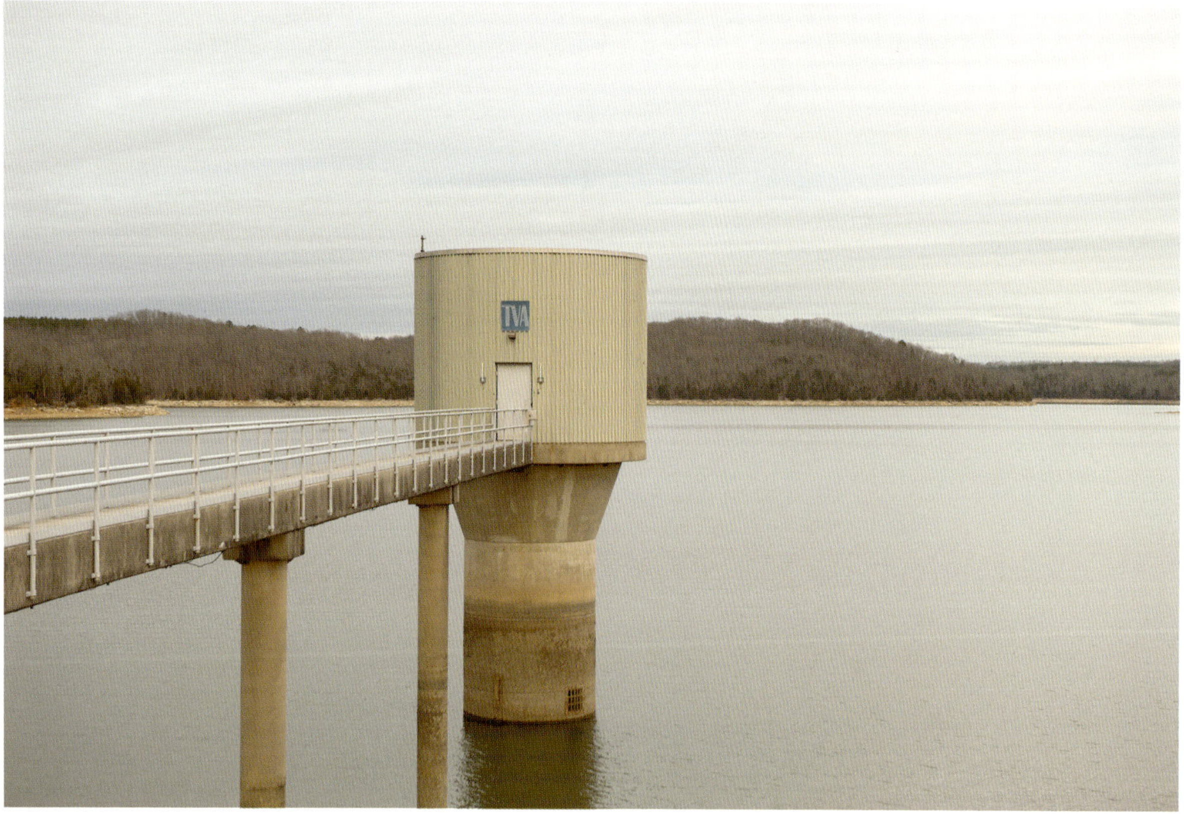

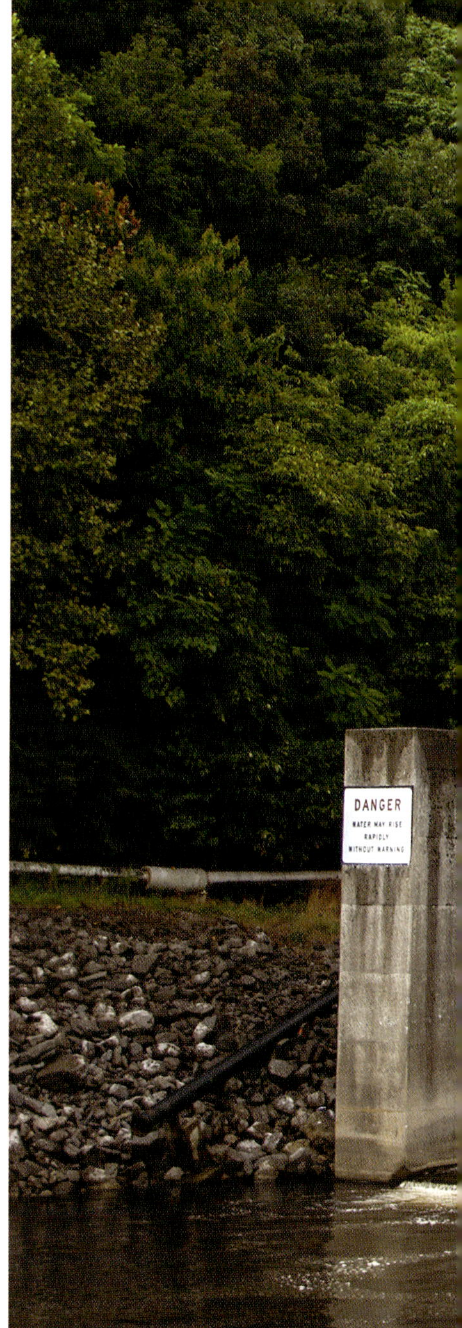

Little Bear Creek, Reservoir

(facing) Normandy

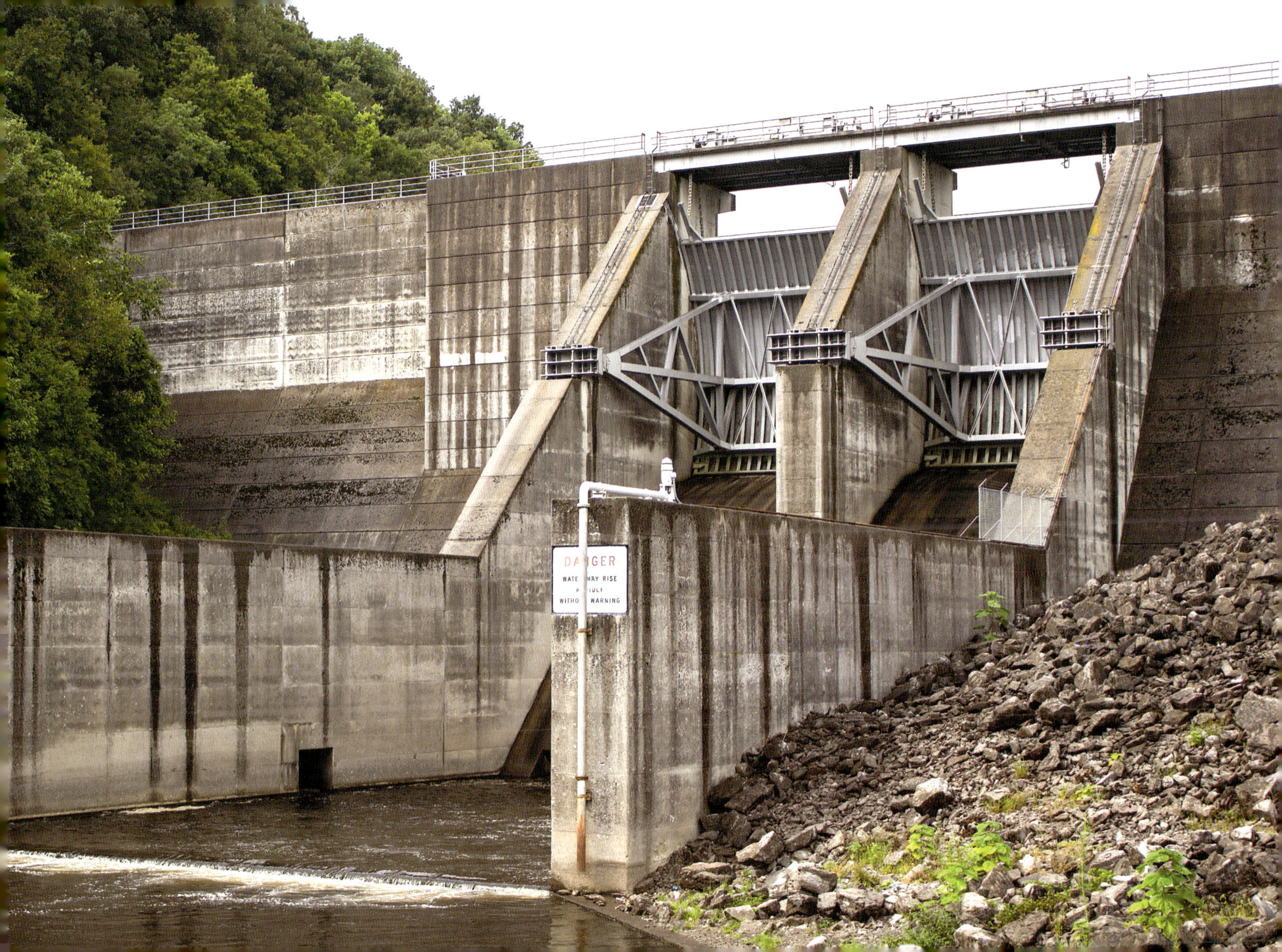
DANGER
WATER MAY RISE
RAPIDLY
WITHOUT WARNING

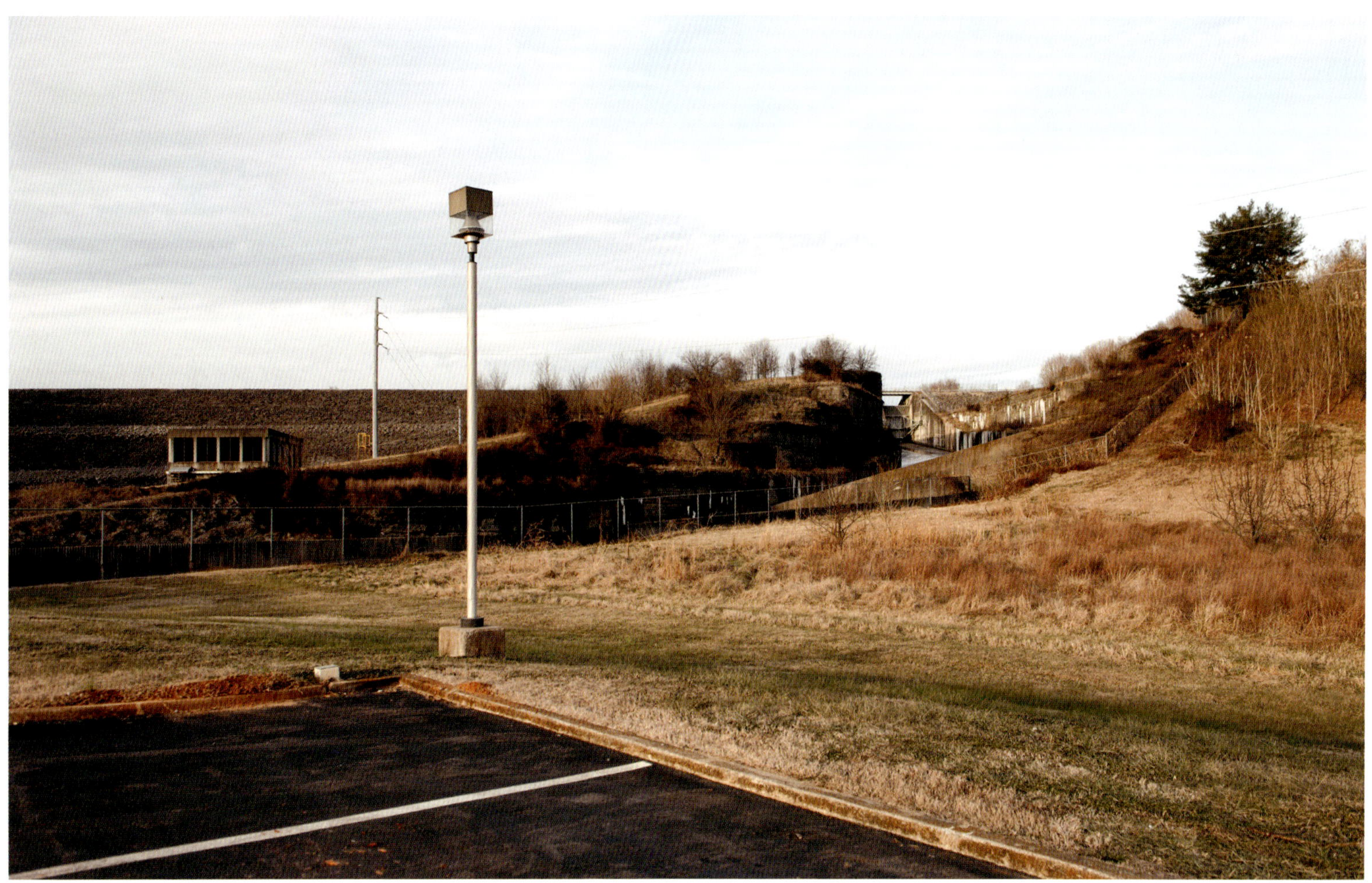

Tims Ford

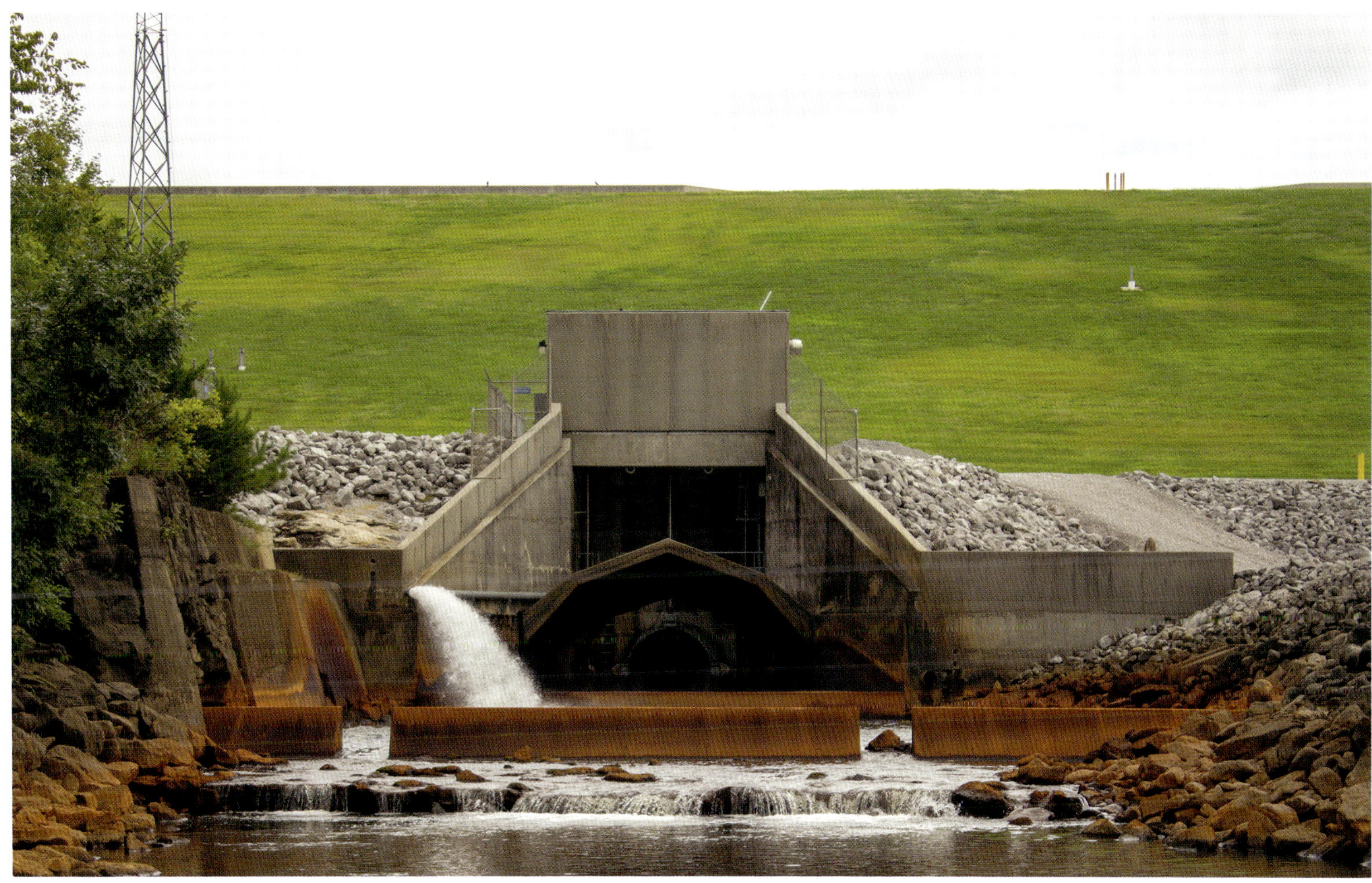

Cedar Creek

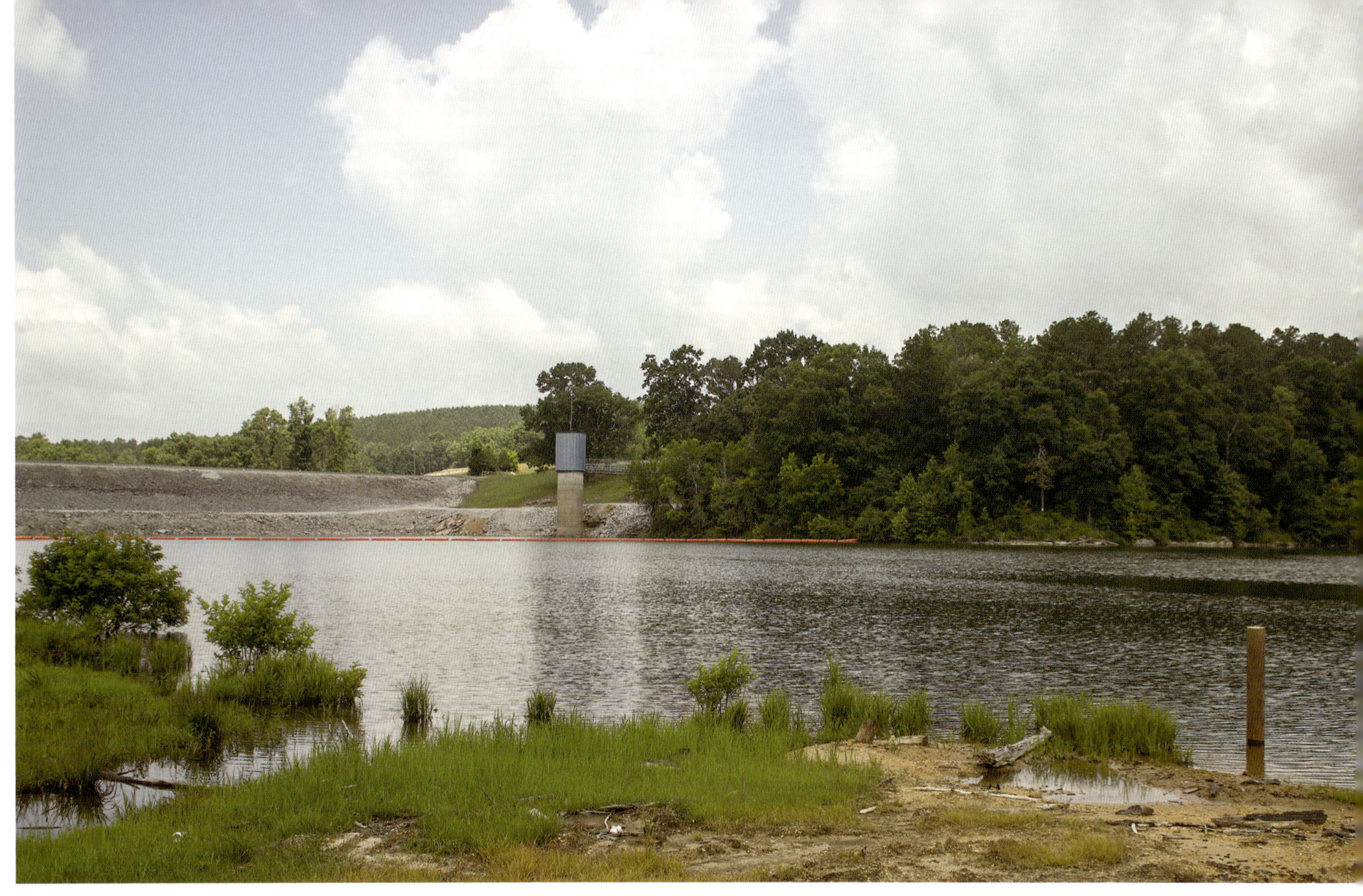

Bear Creek, Recreation Area

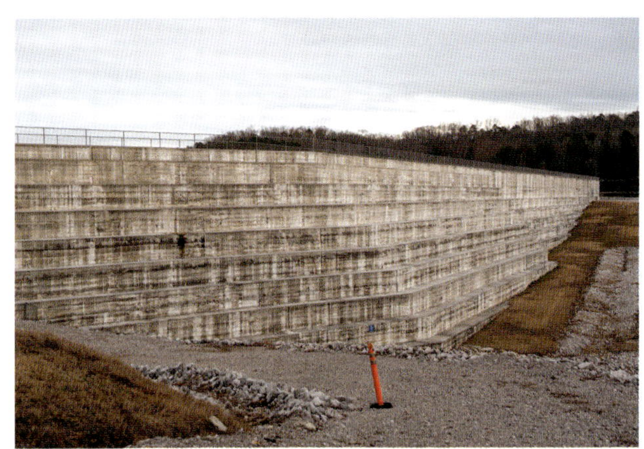

Bear Creek

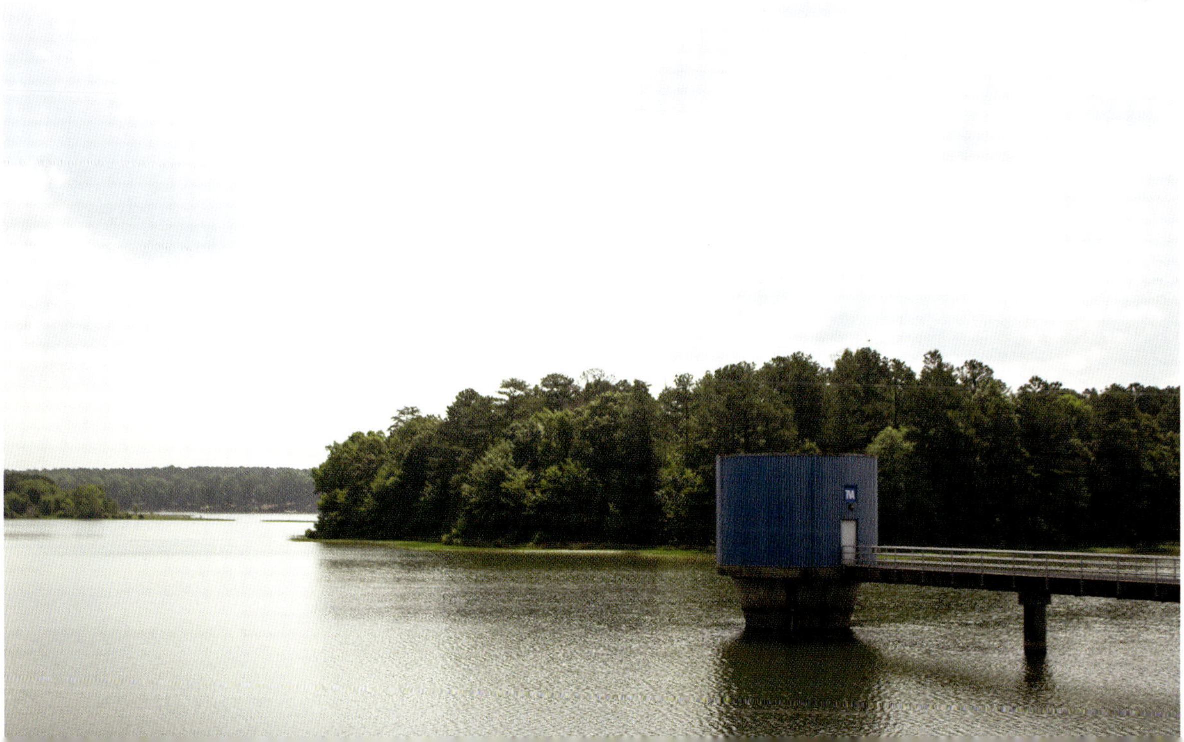

Upper Bear Creek, Reservoir

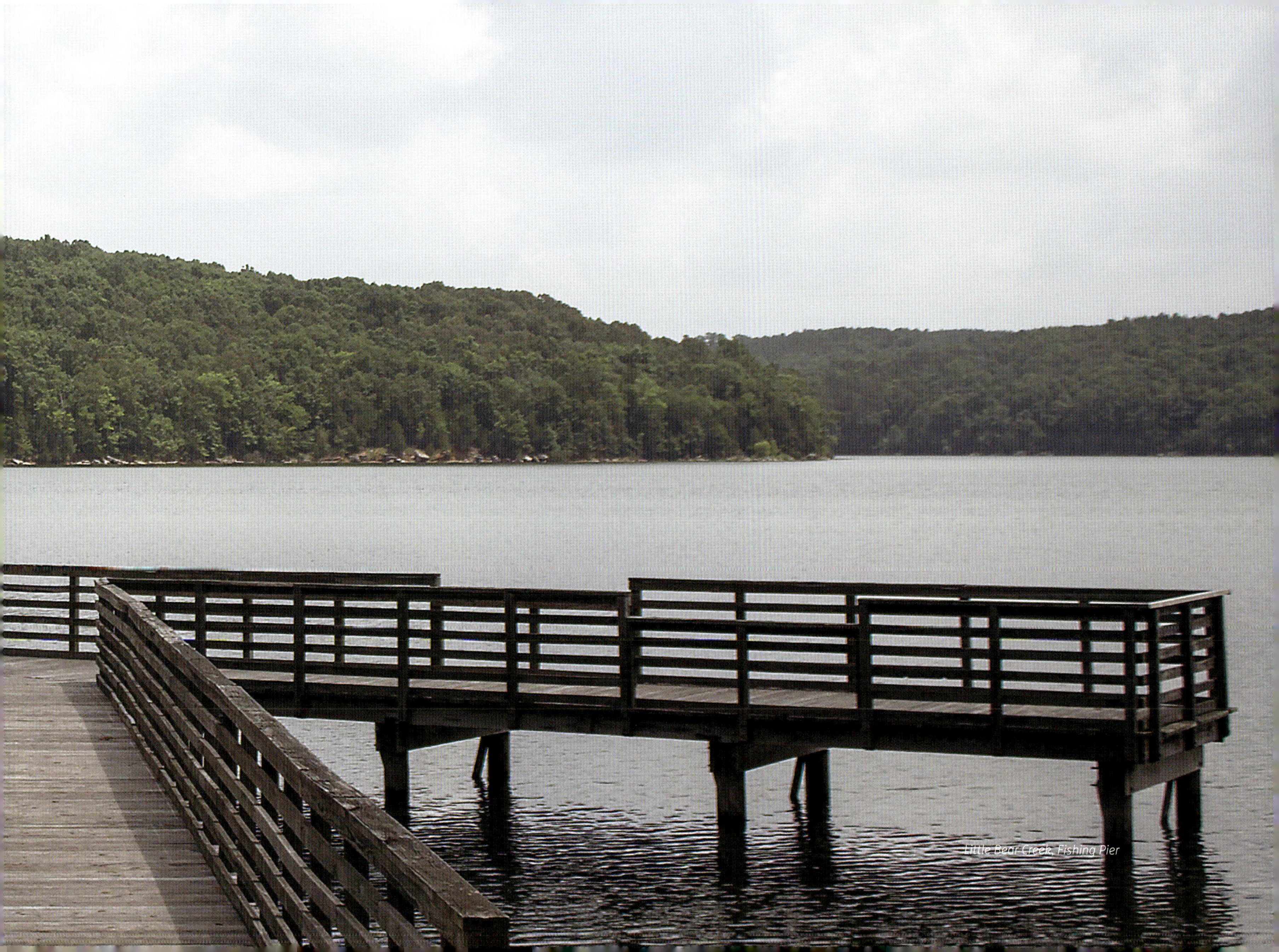
Little Bear Creek, Fishing Pier

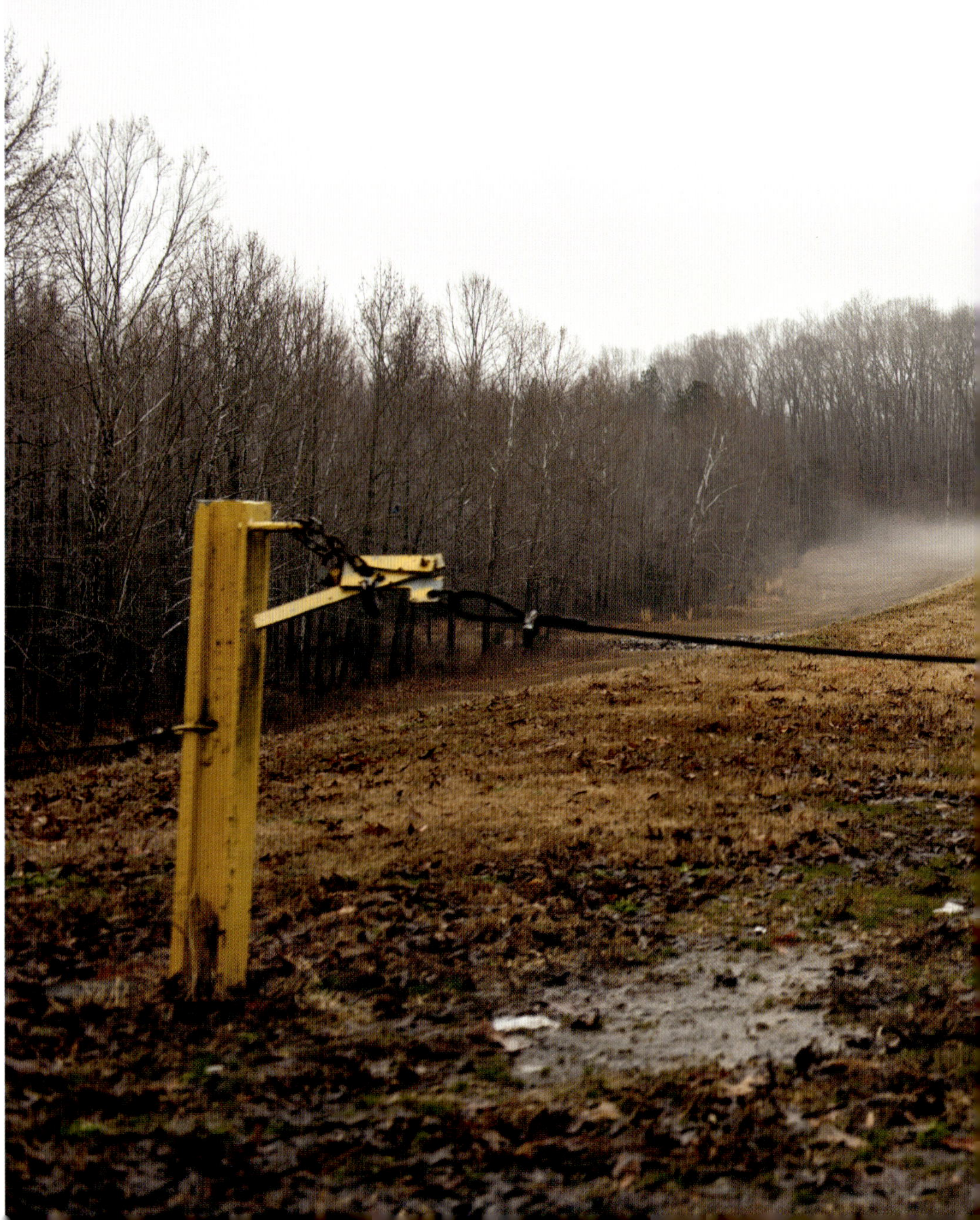

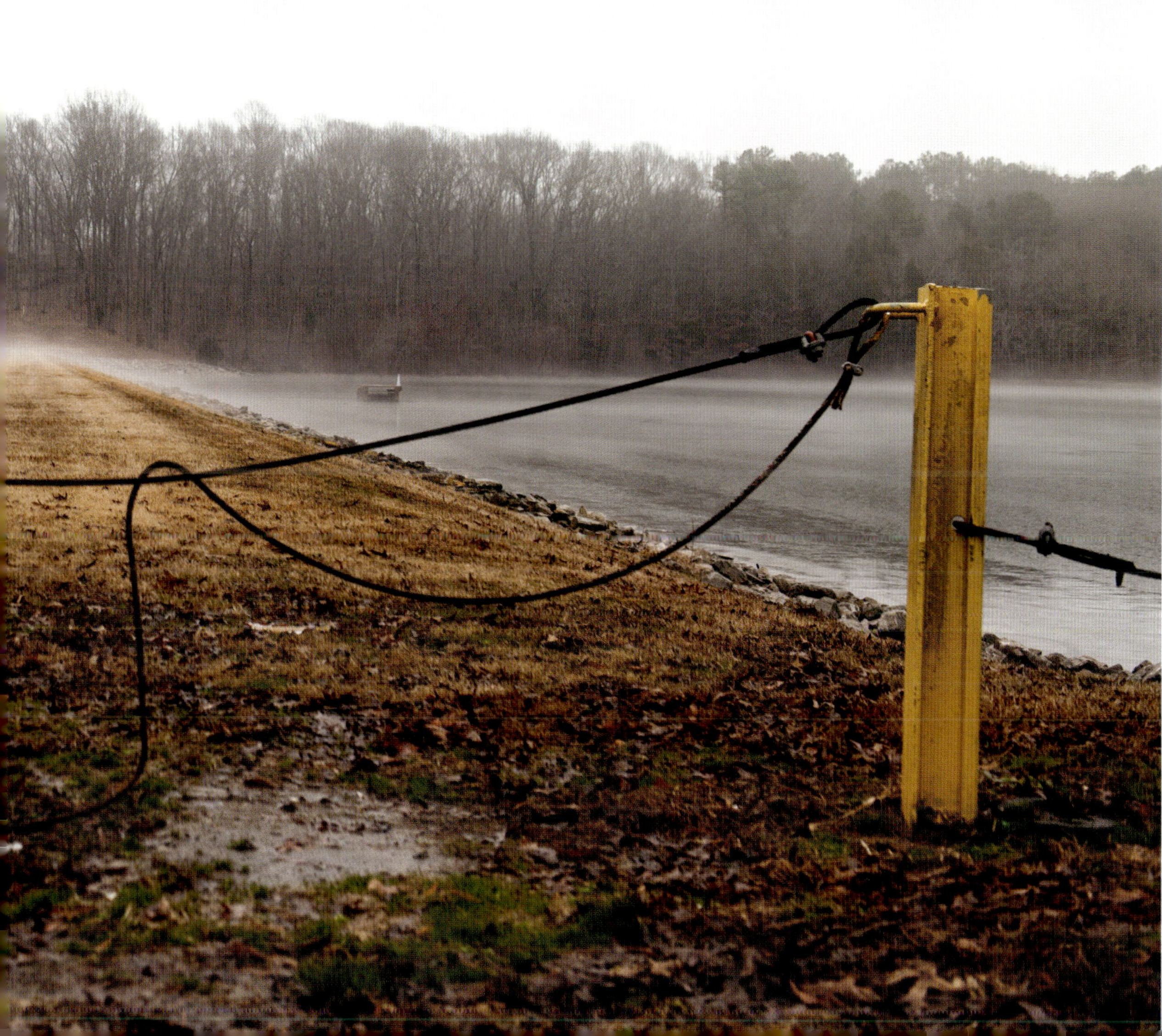

Dogwood

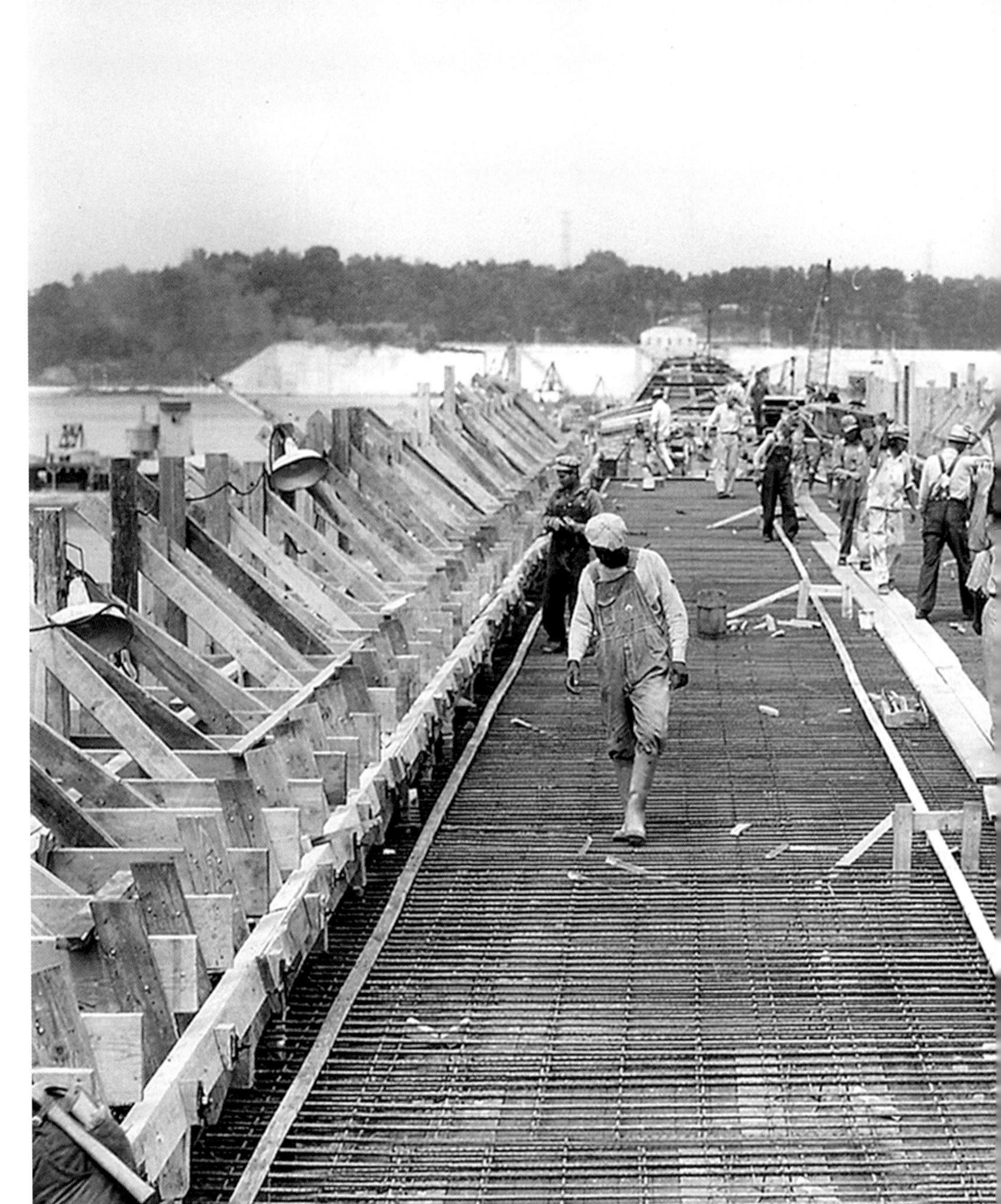

Wheeler Dam (1935) / Tennessee Valley Authority

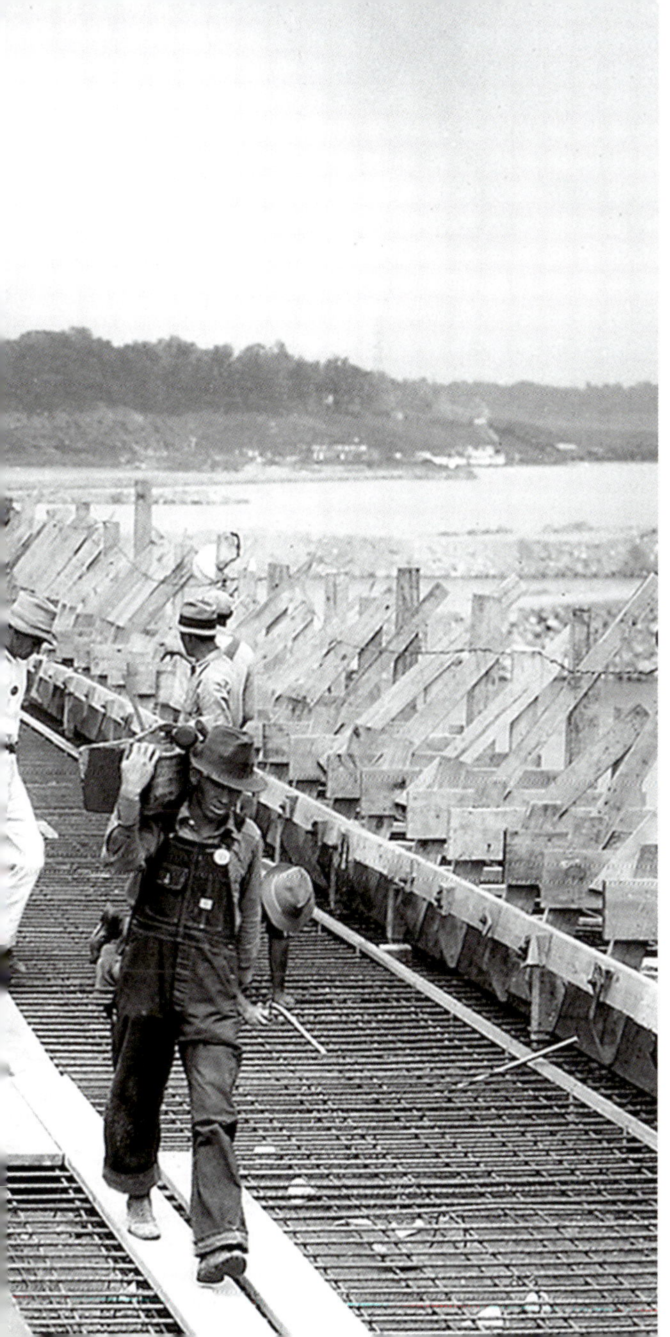

SECTION FOUR

Tennessee River

The main river dams of the Tennessee Valley Authority were built first and foremost for hydroelectricity and reliable navigation. Strikingly industrial, they command respect in a deliberate way and are active elements of interstate commerce. Trucks and automobiles traverse many of the dams, railroad bridges and overpasses sever the surrounding vistas, and the navigation locks' concrete walls stretch out from the dams. The narrative of these dams is utility, yet each location maintains TVA's commitment to public access and recreational opportunity.

For the visitor, the overall square acreage of the Tennessee River dams provides a vastly different experience than what is found along the tributaries. Opportunities for public gatherings are numerous and can be found throughout each reservation. Substantial picnic areas replete with playgrounds, pavilions, and camping spaces are situated both downstream and along the reservoir, and large concrete piers are routinely provided for fishing.

The industrial backdrop of these spaces includes additional visual obstacles, however: power lines, increased boat traffic (both commercial and recreational), and extensive fencing and barriers. The delineation of public and private space is more pronounced at these main river dams than it is at tributary dams: The warning signage is larger and more abundant, and the inherent risk within these locations is more palpable. Even though the powerhouses are closed to the public, walkways and parking lots lead the visitor to the base of the dams where the generation of hydroelectricity can be witnessed, heard, and felt.

While architecturally similar, each main river dam possesses unique visual elements. Fort Loudoun's green bridge crests over the river elegantly, while Watts Bar's shuttered visitor center overlooks the dam's powerhouse. Chickamauga's navigation lock is under repair and surrounded by rust-colored cofferdams, and Nickajack's harshly angular powerhouse overpowers the horizontal lines of its spill gates. Wheeler blocks the horizon like a blackened sea wall, its industrial appearance a direct contrast to Guntersville which is nestled rather peaceably in Alabama's Appalachian foothills. Wilson's iconic neoclassical architecture pre-dates TVA, and

Pickwick Landing's highway bridge stands tall above its reservation, rising upward as it bypasses the dam's switchyard. Finally, Kentucky impounds the entirety of the Tennessee Valley watershed, its massive gantry cranes silhouetted against the sun. The Tennessee River is also home to Raccoon Mountain Pumped-Storage Plant, an engineering marvel that is one of TVA most architecturally interesting and aesthetically pleasing locations. You'll also find the remnants of Hale's Bar Dam, a location that was inherited by TVA in 1933 and decommissioned in the late 1960s.

Given the variety in dam and facilities design at TVA locations along the Tennessee River, I was struck by the one thing they all had in common: the consistent prominence and dominance of "Dangerous Waters" and other warning signs. *Fort Loudoun, Downstream* (page 92) reveals this intense and uncomfortable interaction of visual language and industrial landscape design. *Fort Loudoun, River Access,* (page 100) focuses on a different type of warning, one that underscores the location's ecological instability and undermines recreational usage.

Pickwick Landing, Downstream (page 93) emphasizes the tension between industry and recreation and considers the disjointed nature of the hydroelectric landscape. It is a mistrustful image–literal in its disregard for the message found on the other side of the sign ("Dangerous Waters. Violent Surges Occur Suddenly. Keep Out.") and metaphorical in its contemplation of the long-term viability of TVA's social goals.

The relationship TVA originally intended to nurture between itself and the people it serves is evident in the architecture at each dam reservation, but visitors now experience moments of physical and visual inaccessibility, such as gated walkways, barred architecture, and closed public facilities. *Watts Bar, Visitor Center* (page 100) is a stark example of this conflicting reality. Embedded in the western embankment above Watts Bar Dam, the visitor center is easily visible from the highway. Architecturally, it certainly resembles a visitor center, with a circular design and exterior guard rails that allow people to take in the powerhouse and dam below. Upon closer examination, it appears accessible by a pedestrian bridge that connects it to a small parking area across the highway, but the photograph taken clearly illustrates the disappointment that greets the curious tourist.

Similar visuals are found at Wilson Dam. Signs along the highway direct motorists to the facility's visitor center, yet no such facility exists. Visitors pull into a small parking lot only to find a few didactic panels outlining the history of the dam and TVA's current energy and environmental initiatives. *Wilson, Visitor Center* (page 101) captures the only acknowledgment that there ever was a visitor center at all, as the dam's powerhouse and its formerly public facility are guarded behind the security fence.

I do not believe that the inaccessibility and inhospitable symbolism suggested by these two photographs is malicious, nor do I believe that the agency is inconsiderate or secretive. It is the opposite: TVA's publicly owned dams are so well designed and inclusive of visitors that it is jarring when sightlines are obstructed by barriers, fences, and security gates. In addition, many of the visitor centers and powerhouses were closed due not only to logistical difficulties and staffing problems but also due to potential security risks. Nonetheless, I decided to include photos such as *Watts Bar, Visitor Center* and *Wilson, Visitor Center* in this book because I believe TVA can do better. If I am able to stand inside Kentucky Dam's visitor center and view the dam's control room, certainly similar opportunities can be found for locations as historically significant and as architecturally beautiful as Watts Bar or Wilson. I am encouraged by the reopening of the buildings at South Holston and Fort Patrick Henry and know first-hand that TVA understands the power of public perception. Its commitment to communicating the importance of these dams to the region is genuine as is its obligation toward public access.

At this time, Wilson Dam does not have a true visitor facility, but it does provide substantial river and lake recreation areas. Locations such as these can offer plentiful opportunities for picnicking, fishing, and all-around outdoor recreation. The most recently updated location at Wilson Dam is the outstanding Rockpile Recreation Area. The space includes a new TVA-themed playground and a pedestrian walkway leading to a collection of waterfalls near the dam's powerhouse. Delightful and clever, the playground features slides that mimic the spillways of the nearby dam, complete with the iconic "Dangerous Waters" warning. Such tongue-in-cheek detail provides hope that the conflicting visuals of shuttered visitor centers and fenced-in powerhouses will be rectified in a thoughtful way.

Many of the photographs in this section define the industrial tonality of these locations from the point of view of the visitor. TVA strives to provide quality amenities in the form of benches, walking paths, fishing piers, overlooks, and so on; but the fact remains that these amenities inhabit landscapes with strong visual reminders of the predominant reason for their construction—hydroelectric power. *Chickamauga* (page 111) and *Nickajack* (page 96) do not shy away from this reality but instead explicitly consider the hydroelectric vistas intended by the agency. Both of these photographs were made from recreational locations with a direct line of sight toward the dams' respective powerhouses. Being in these spaces can be a conflicting experience, whether the agency designed them for "maximum transparency" or not. While the dam reservations routinely support a sense of wonder, discovery, community, and relaxation, there are times when this veneer wears thin. Experiences and visitations within these places are a carefully orchestrated event with preferred sightlines that are emphasized by the landscaping and architecture, while unappealing visuals and mechanisms are routinely minimized.[1]

There is a particular dam that could in no way ever be minimized. Situated twenty-two miles upstream from where the Tennessee River empties into the Ohio, Kentucky Dam is the longest structure in TVA's system at over a mile and a half long and impounds TVA's largest reservoir.[2] The photographs I made on its reservation are in response to its earnestness. *Kentucky* (page 108–109) captures the scale of the dam and its utilization along the river. One might assume that Kentucky Dam's industrial atmosphere could not possibly lend itself to a leisurely afternoon of fishing, yet I saw dozens of anglers who seemed at peace with their engineered surroundings. They showed no concern for the huge boils of water roiling up near the powerhouse or the intensity of the water spilling out of the dam's gates. Instead, they seemed to possess implicit trust in the agency's ability to generate electricity while ensuring their safety.

This kind of trust sustains TVA within the communities it serves. It has been passed down through generations and is on display at the hydroelectric dams seen throughout this book. TVA facilities were born out of sacrifice and promise and were nurtured to provide stability and opportunity. These locations were designed to harness nature but also create accessibility to previously untamed rivers and landscapes. TVA is a company owned by the people, but it also tightly controls the land it took from them. It is true that these dualities foster pockets of resentment and mistrust, and that these negative perceptions were founded in valid grievances. However, in my experience, TVA is appreciated by a significantly large number of citizens. People recognize that they have benefited from the social and cultural changes the company brought to the region, and they are thankful for the dams that continue to play a meaningful role in their lives. This latter group of grateful citizens uphold the spirit of the Tennessee Valley Authority, acknowledging that it was created for the public good and is maintained by the public's trust.

NOTES

1. Todd Smith, "Almost Fully Modern: The TVA's Visual Art Campaign," in *The Tennessee Valley Authority: Design and Persuasion*, ed. Tim Culvahouse (New York: Princeton Architectural Press, 2007), 111–12.
2. "Kentucky," Tennessee Valley Authority, https://www.tva.gov/Energy/Our-Power-System/Hydroelectric/Kentucky-Reservoir.

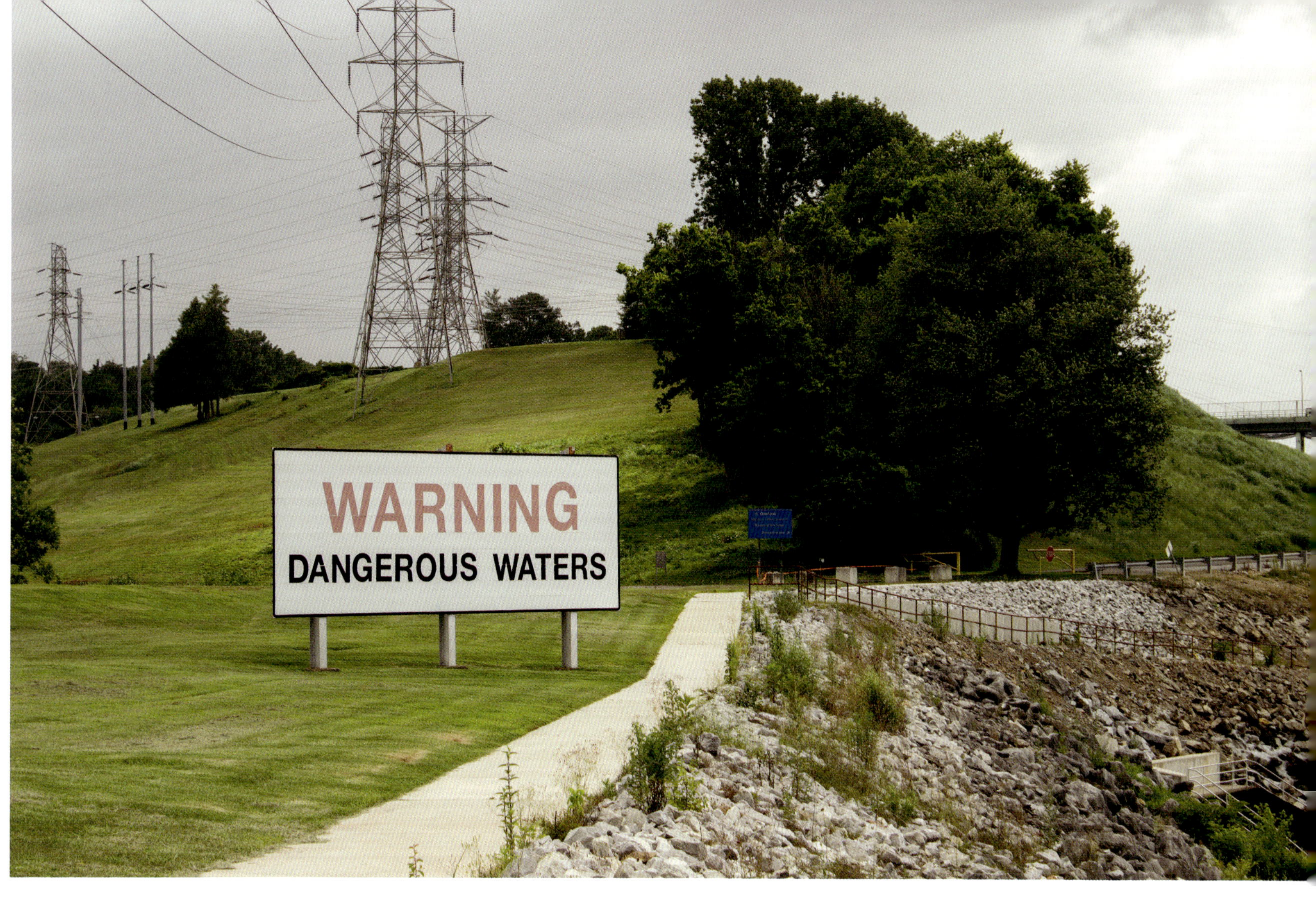
WARNING
DANGEROUS WATERS

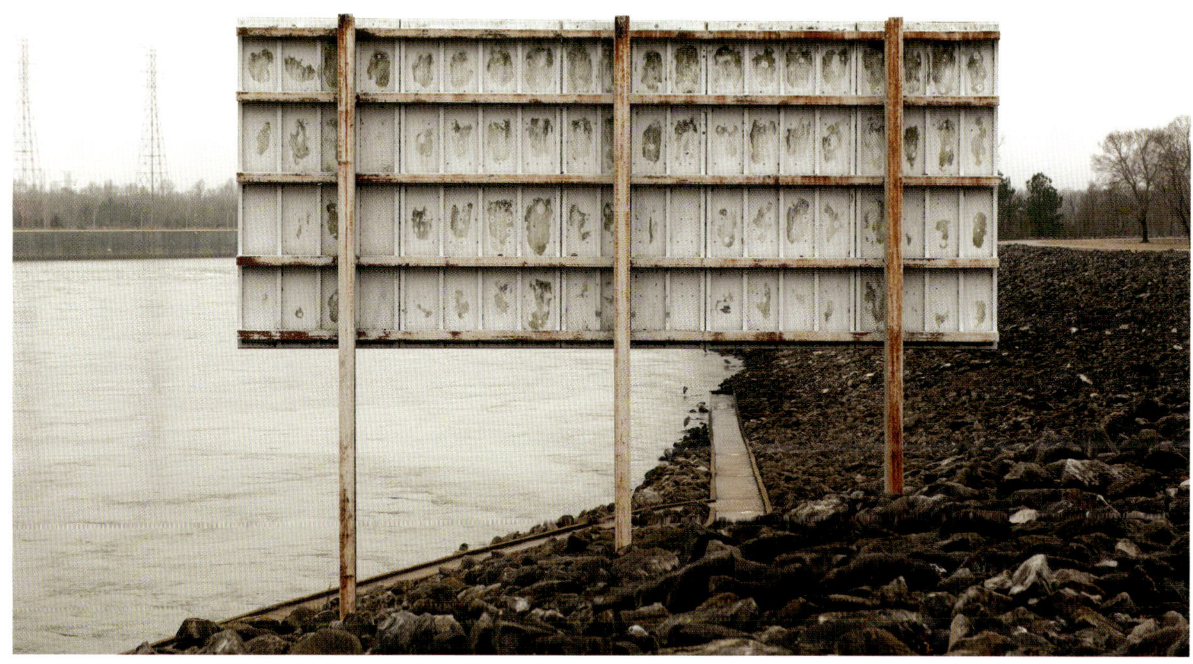

Pickwick Landing, Downstream

(facing) Fort Loudoun, Downstream

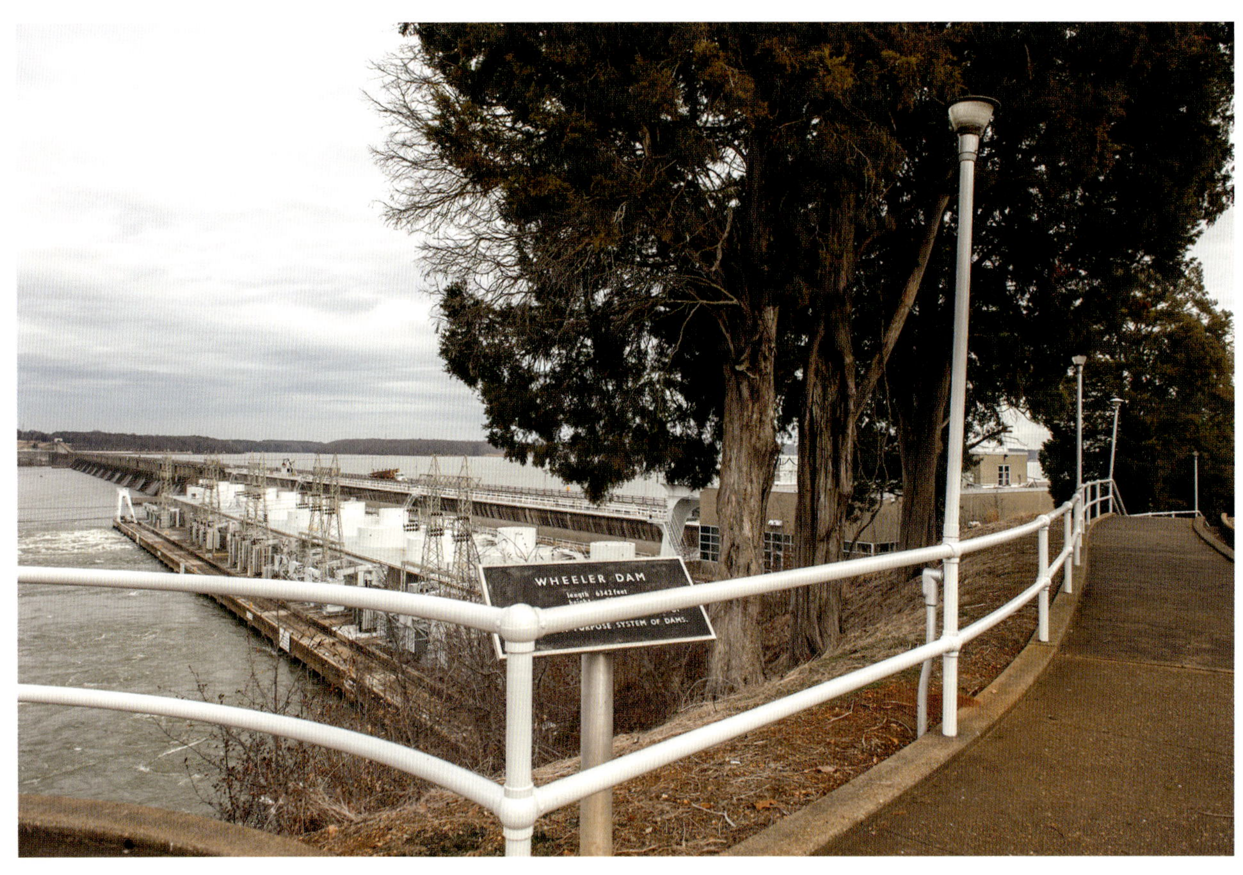

Wheeler

(facing) Raccoon Mountain

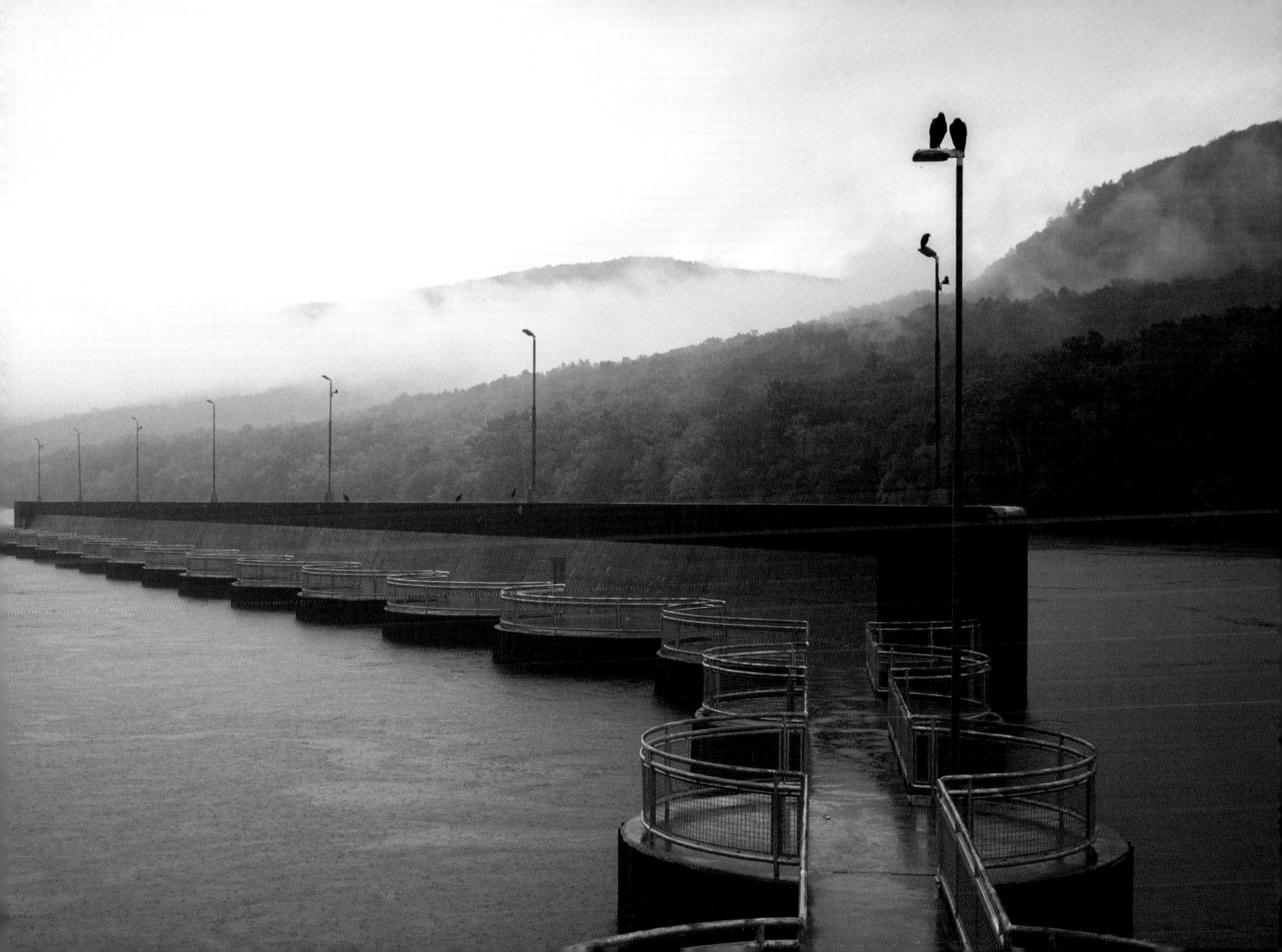

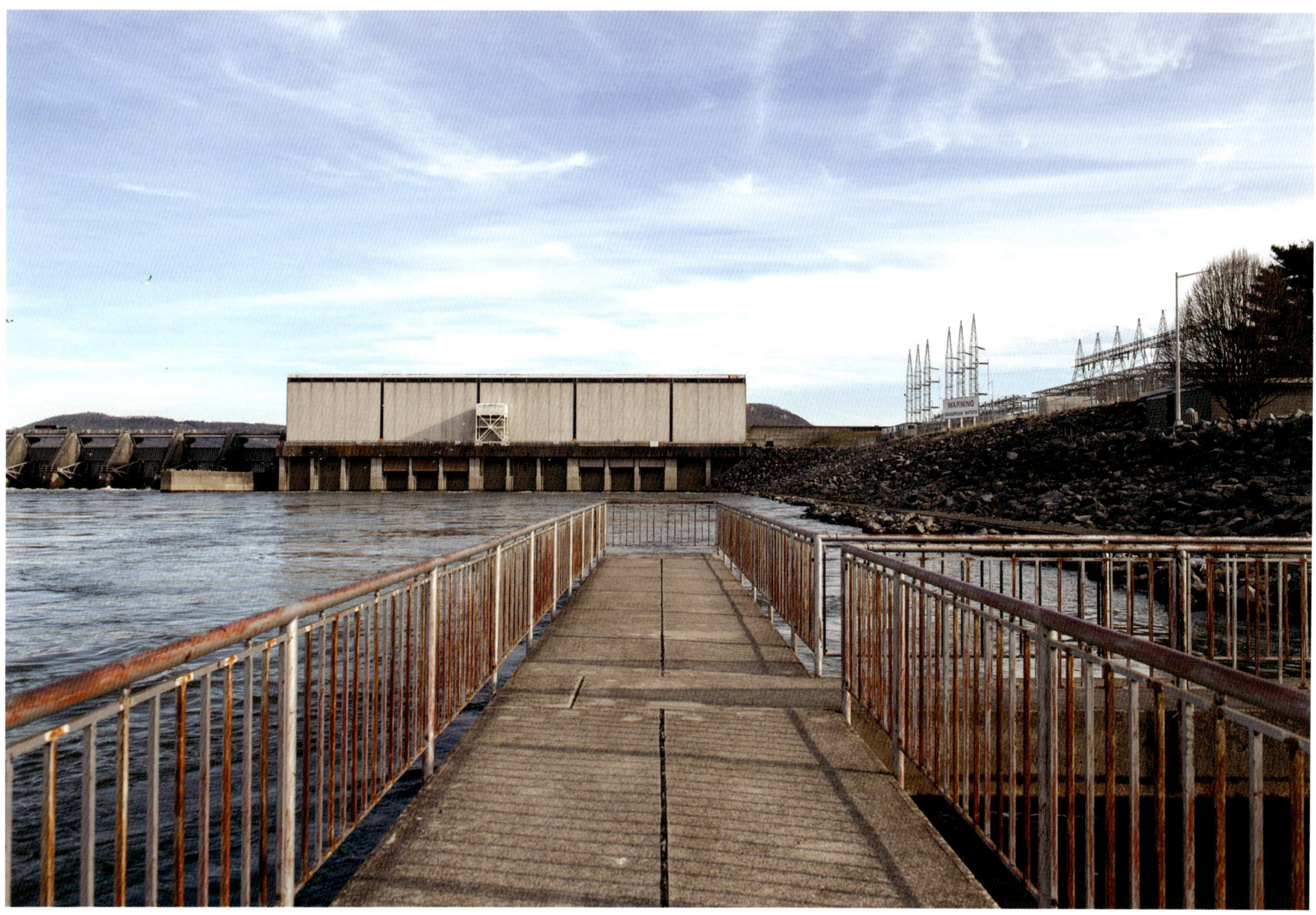

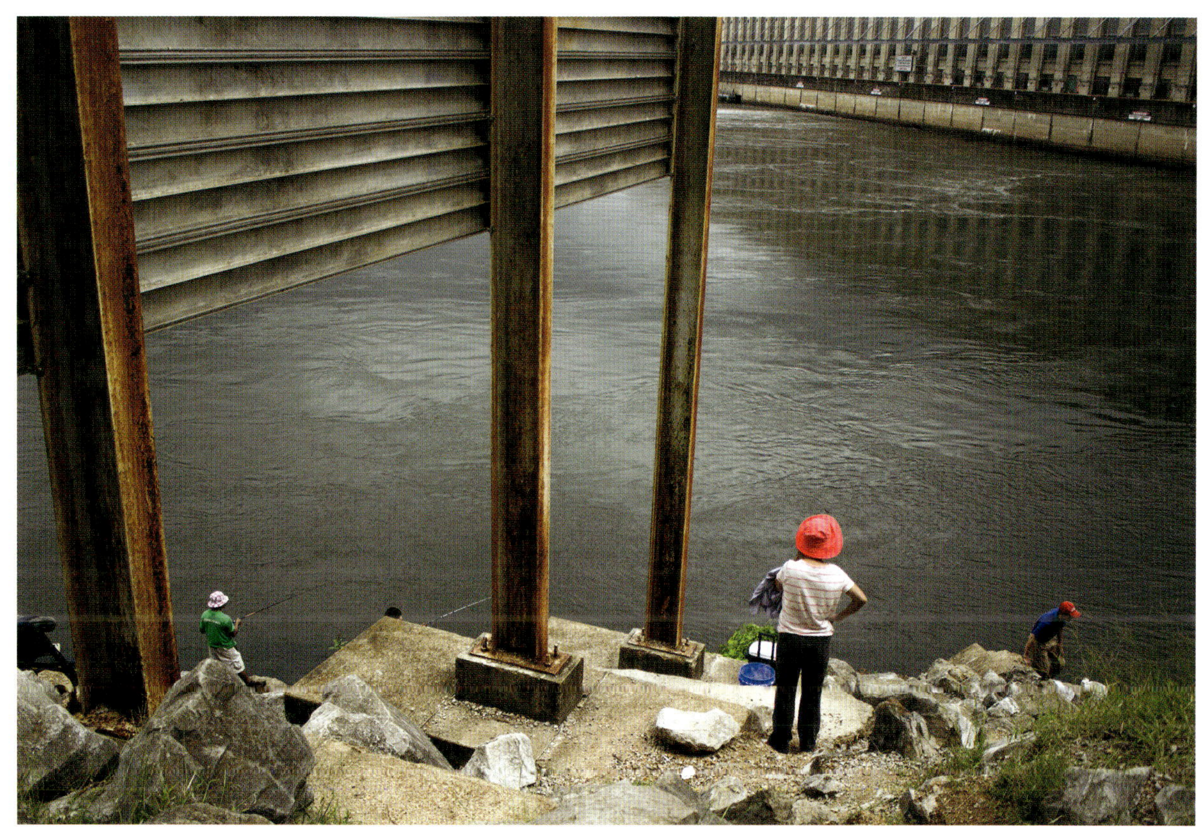

Wilson, Downstream

(facing) Nickajack

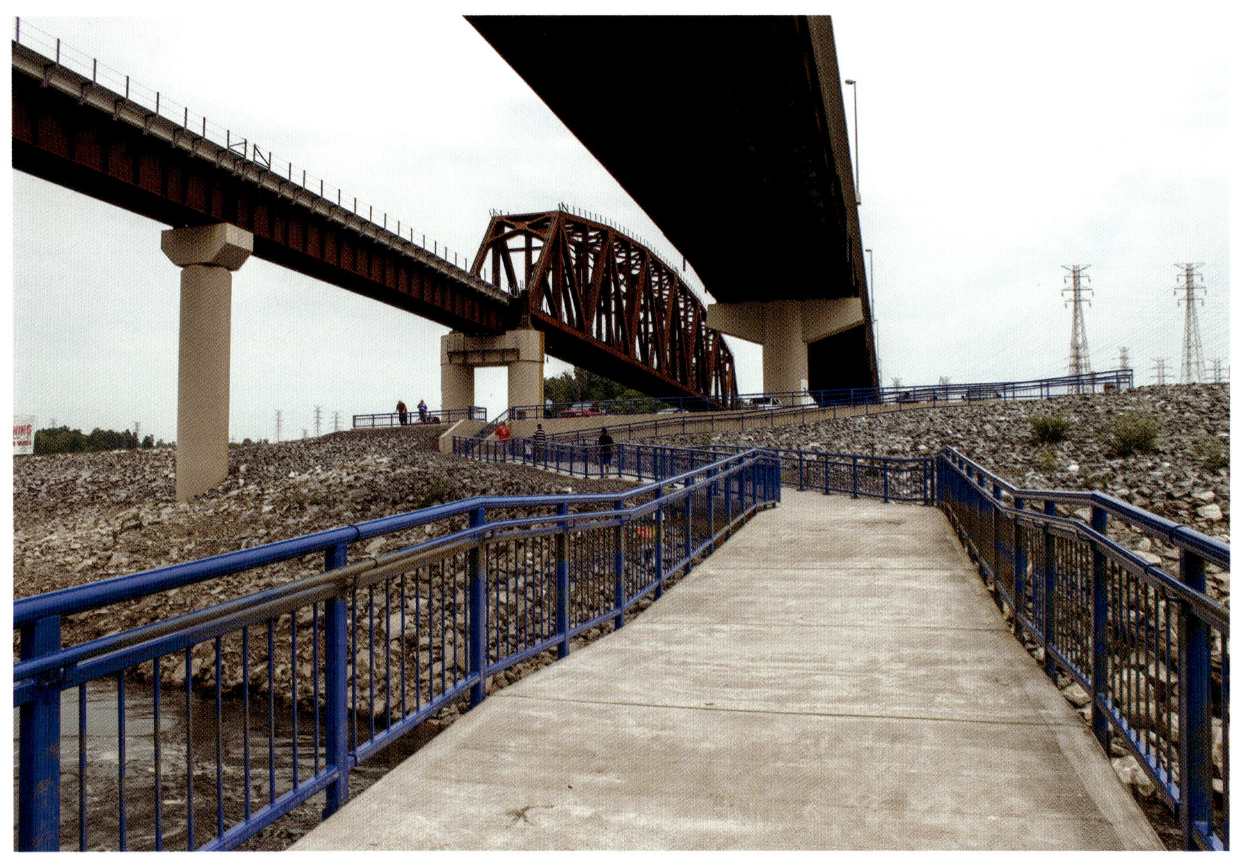

Kentucky, Fishing Pier

(facing) Fort Loudoun, Upstream

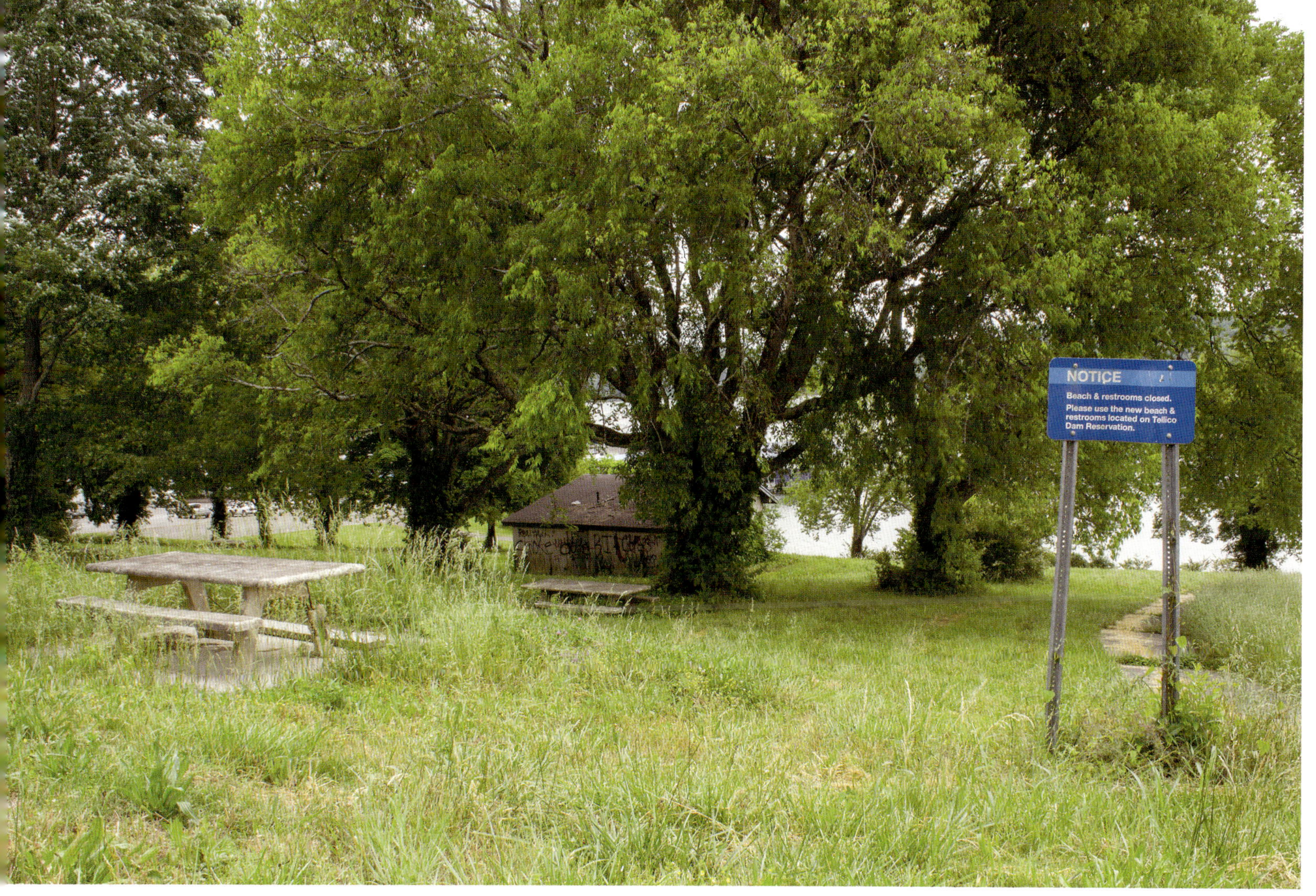
NOTICE
Beach & restrooms closed.
Please use the new beach &
restrooms located on Tellico
Dam Reservation.

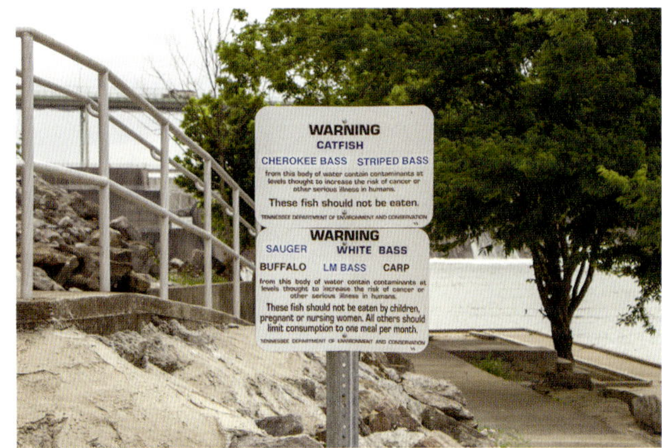

Fort Loudoun, River Access

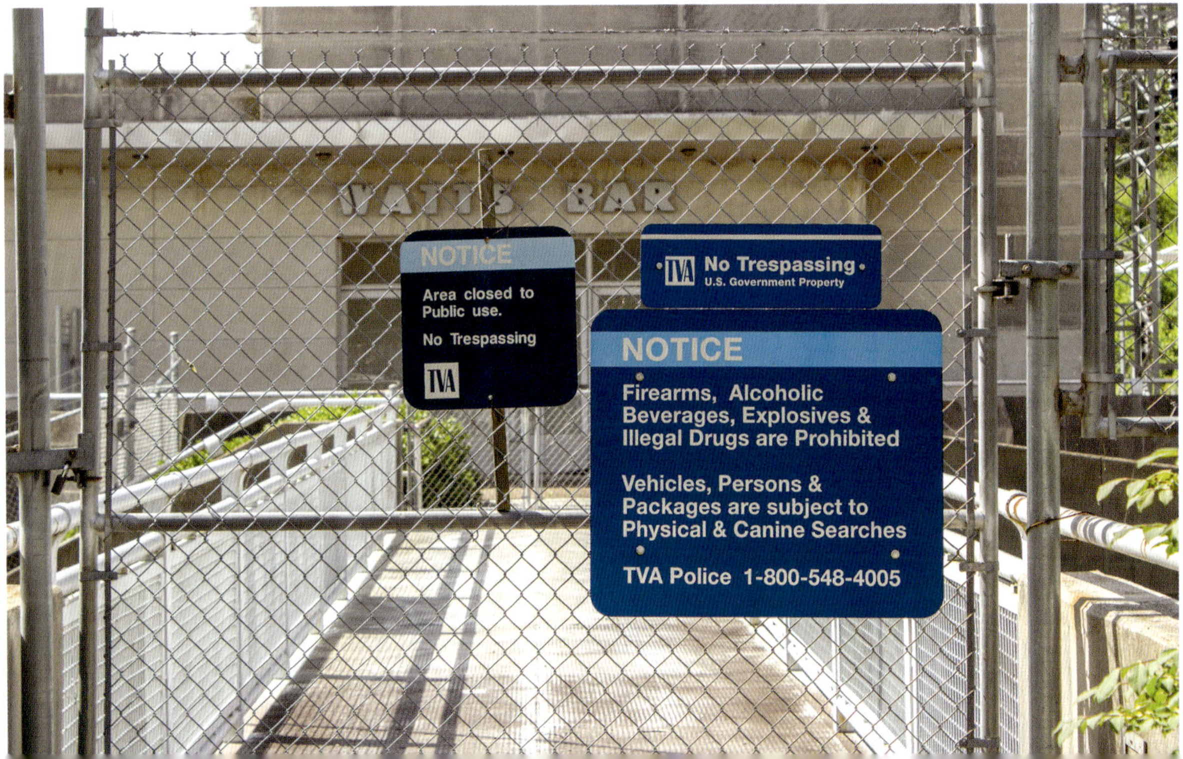

Watts Bar, Visitor Center

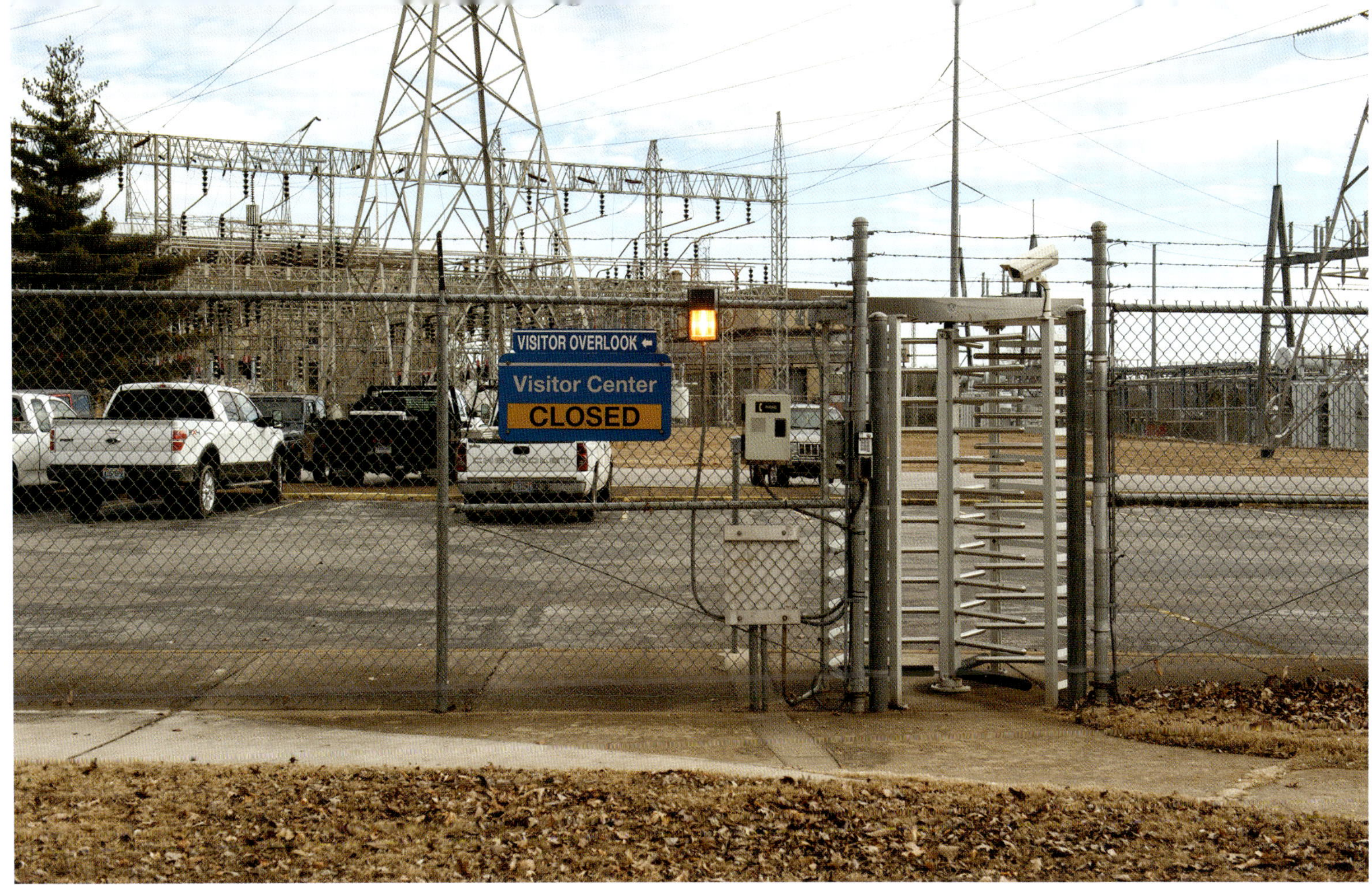

Wilson, Visitor Center

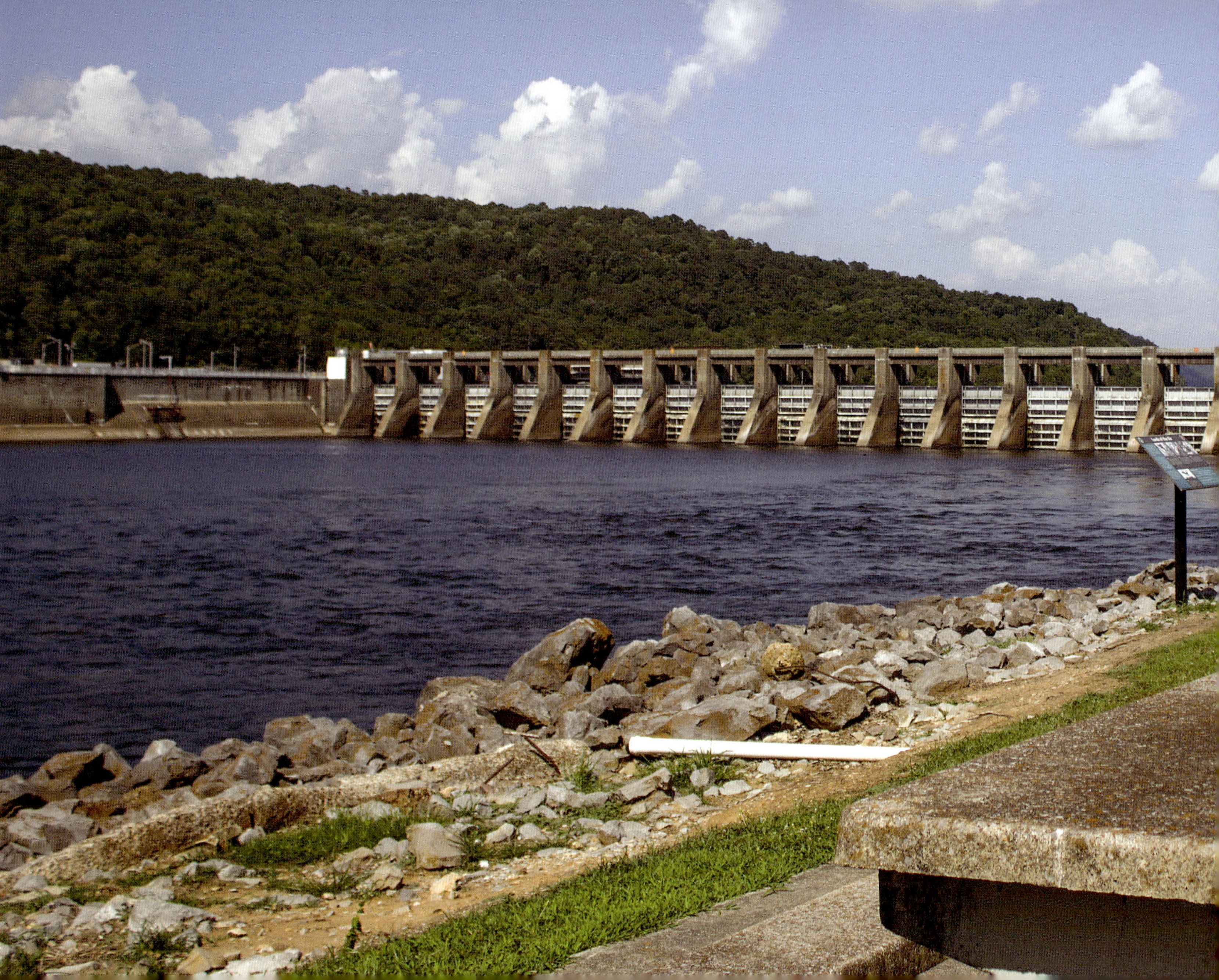

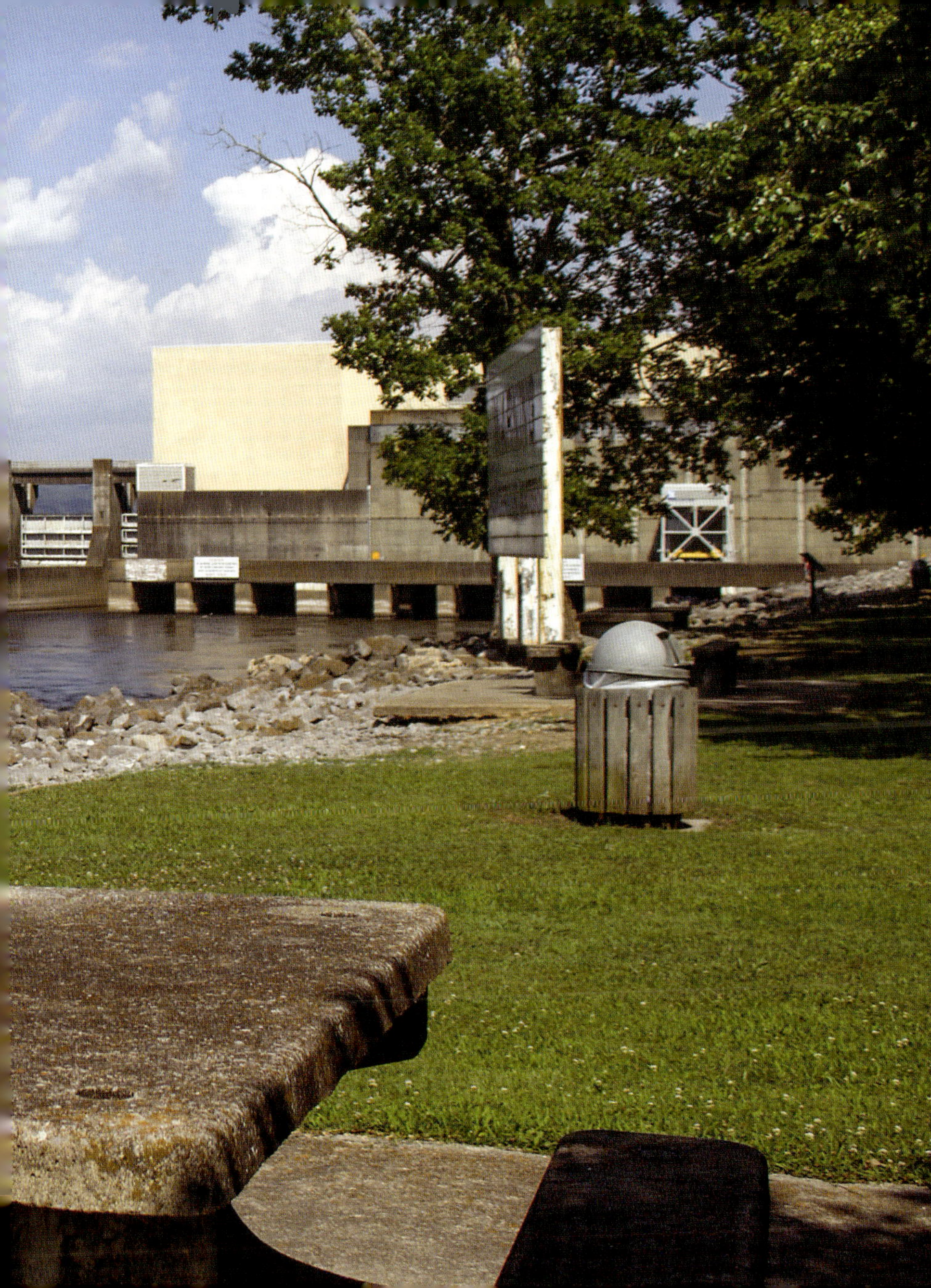

Guntersville

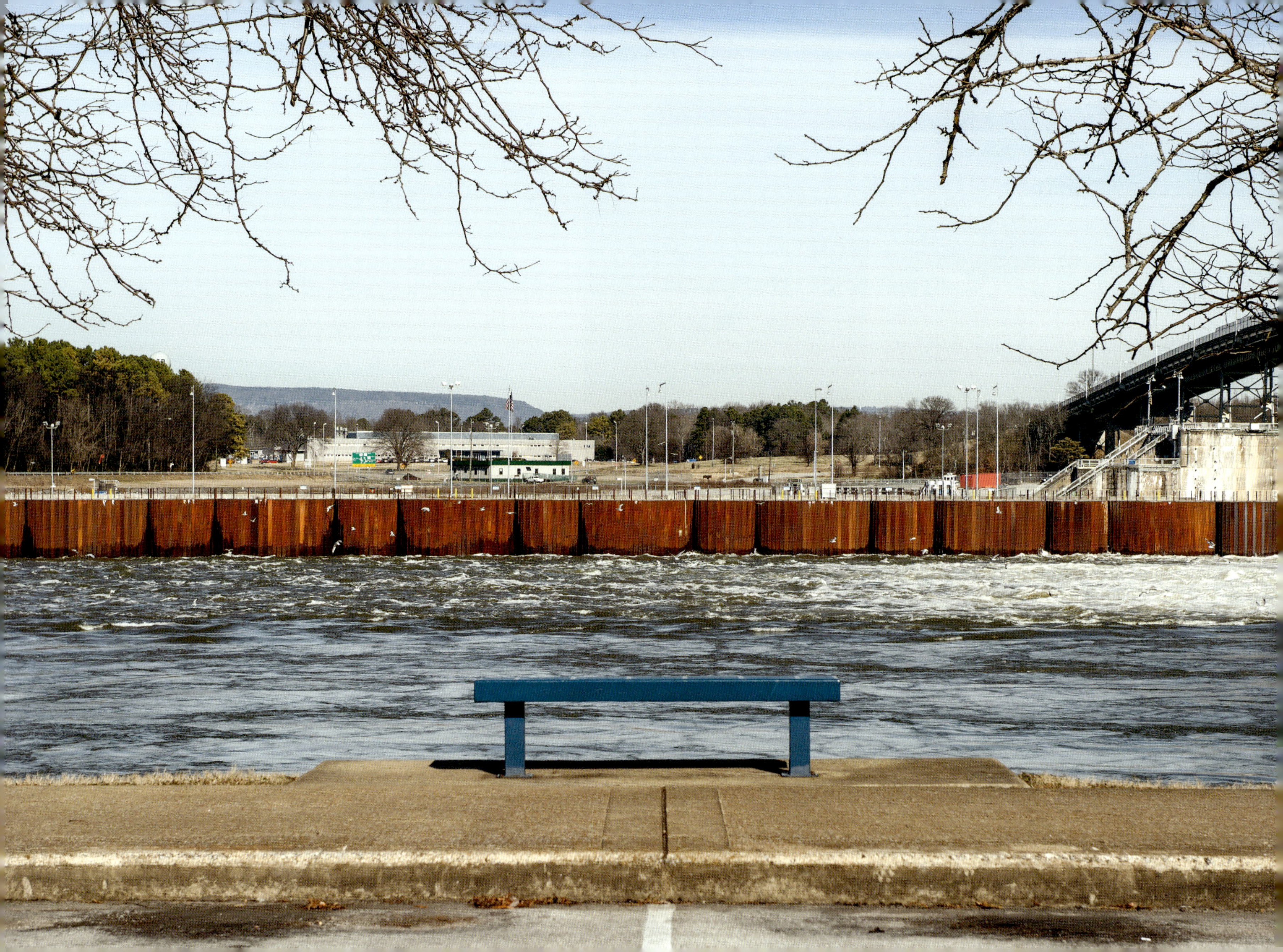

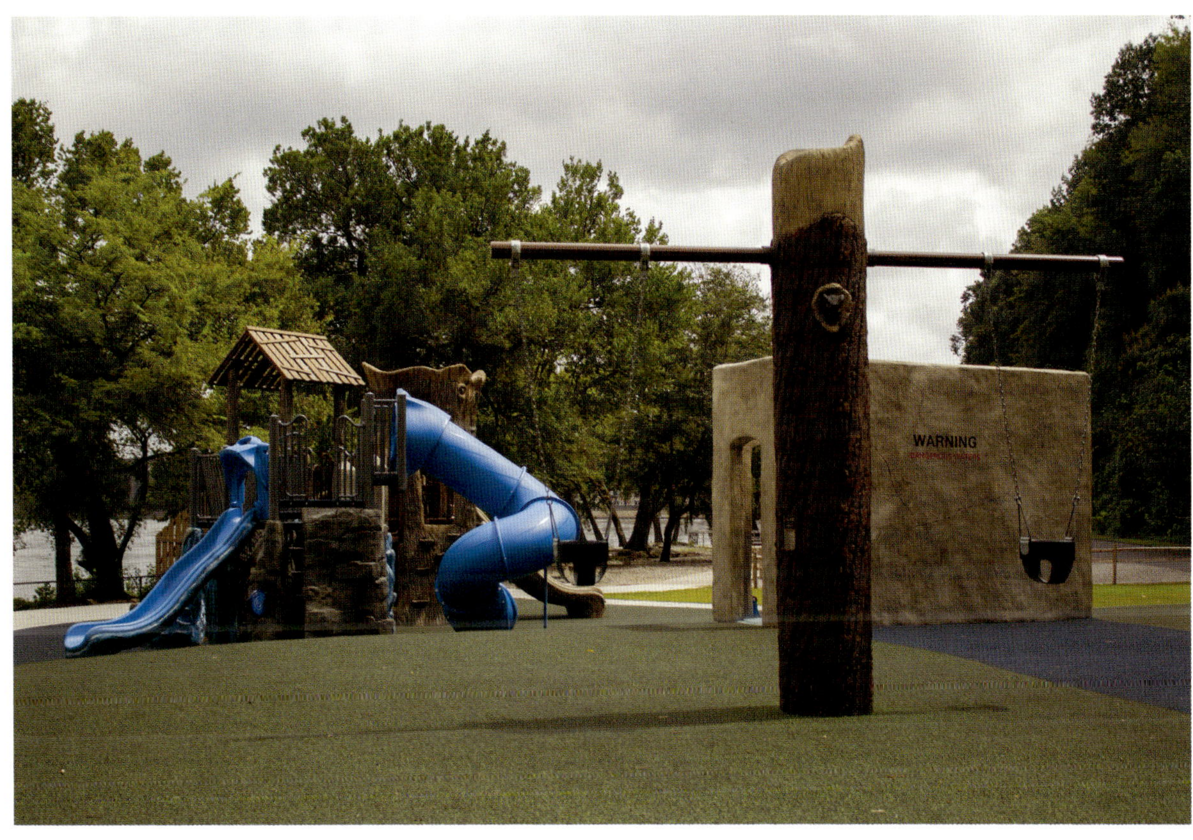

Wilson, Recreation Area

(facing) Chickamauga, Downstream

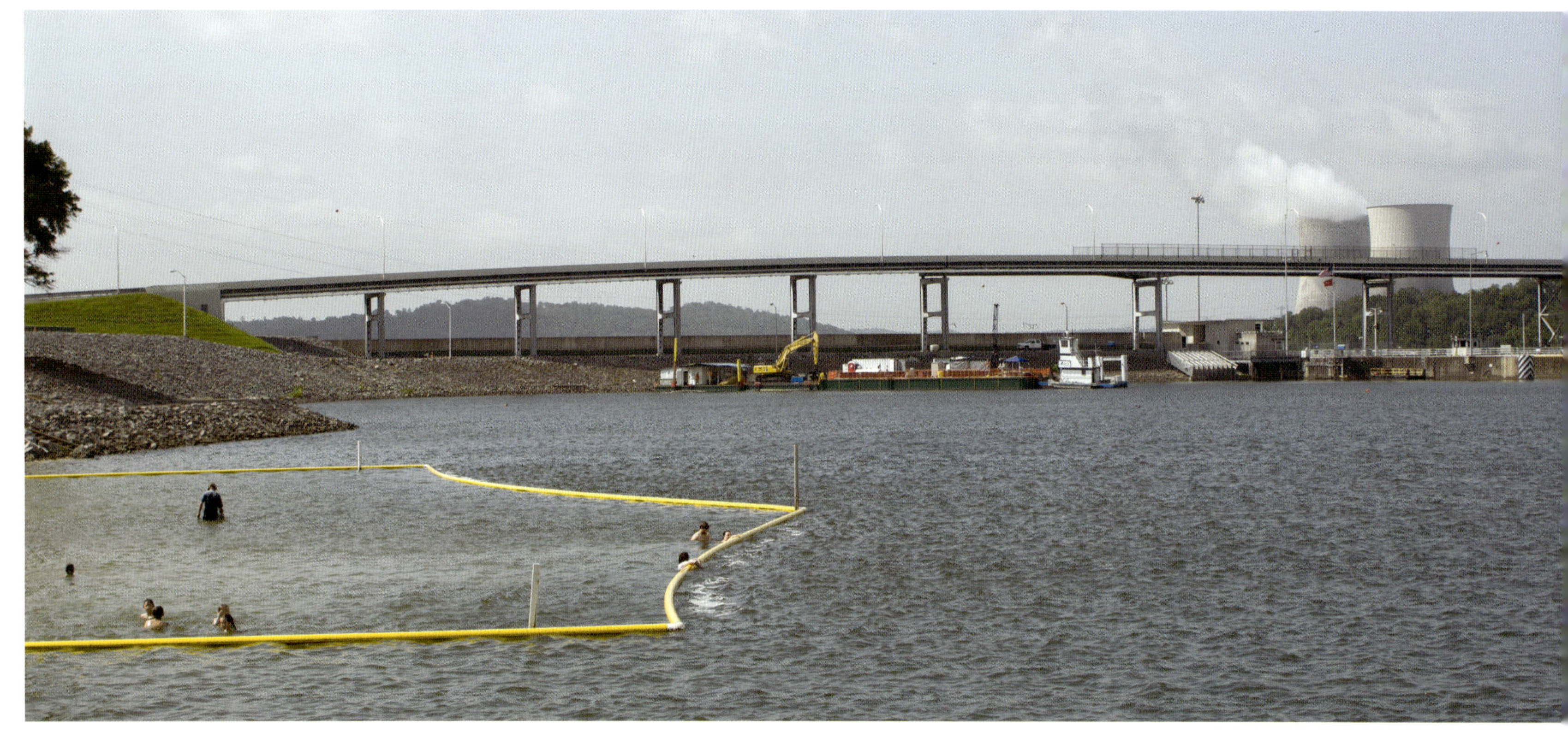

Watts Bar, Recreation Area

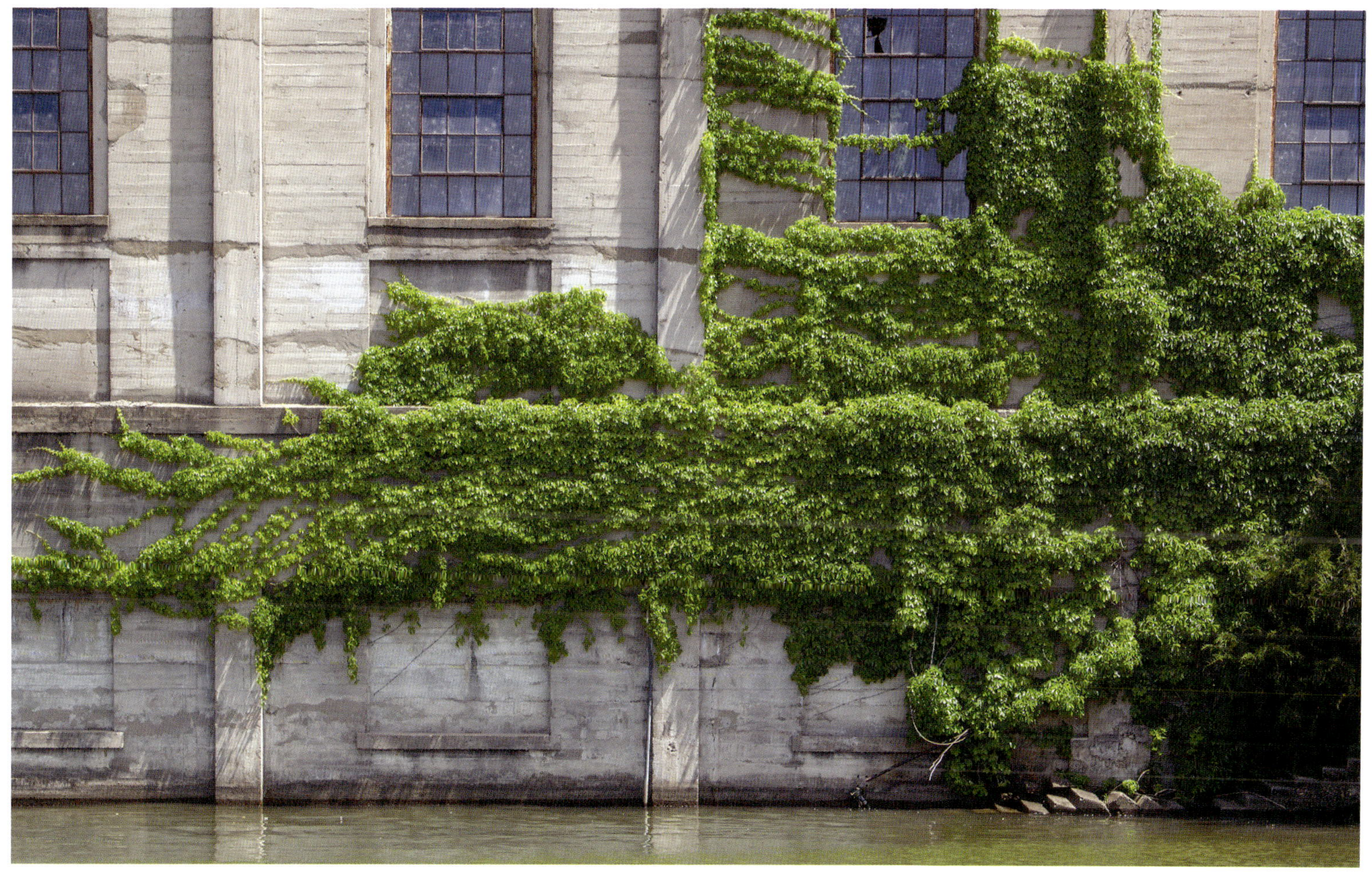

Hales Bar, Powerhouse

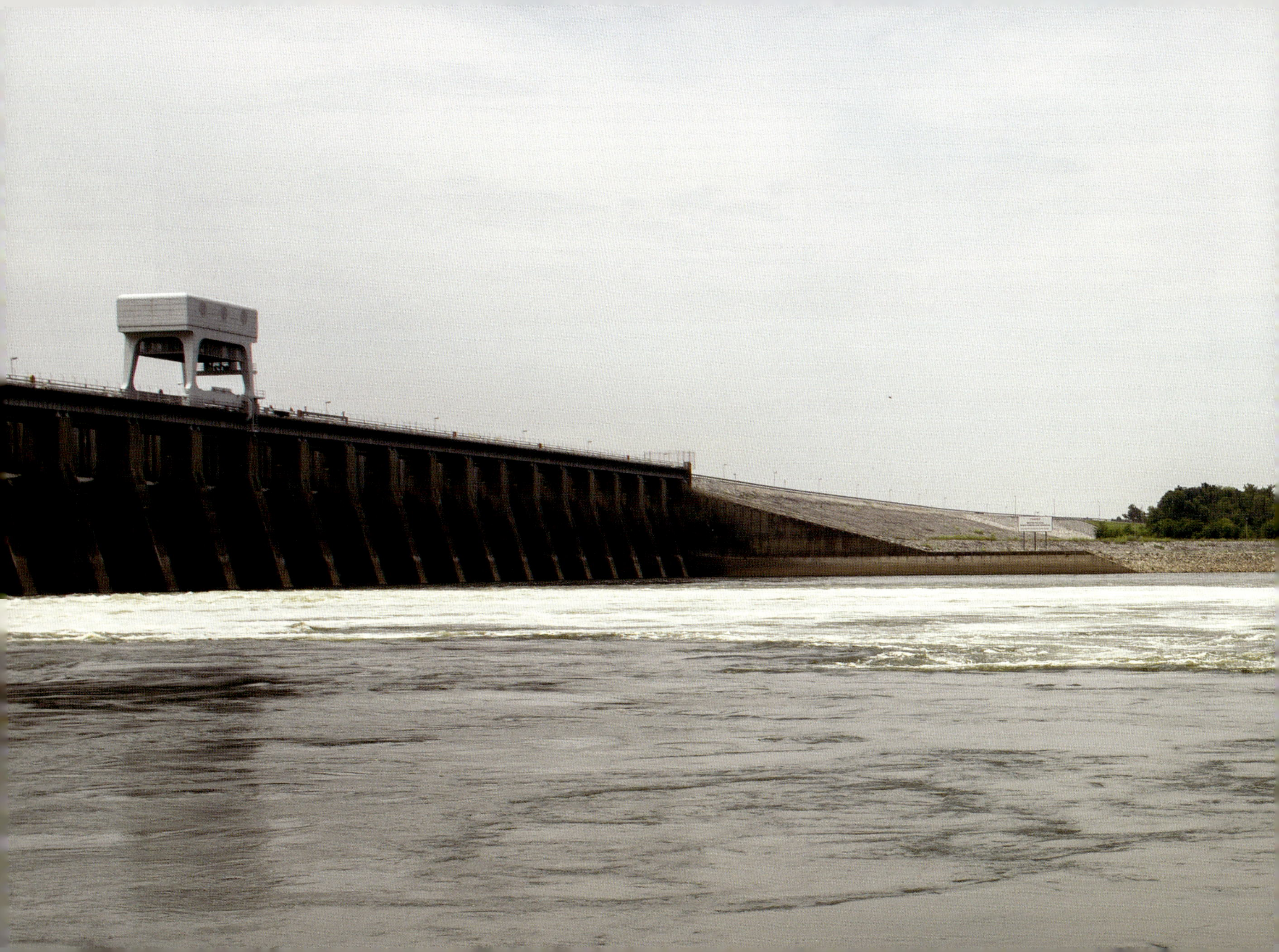

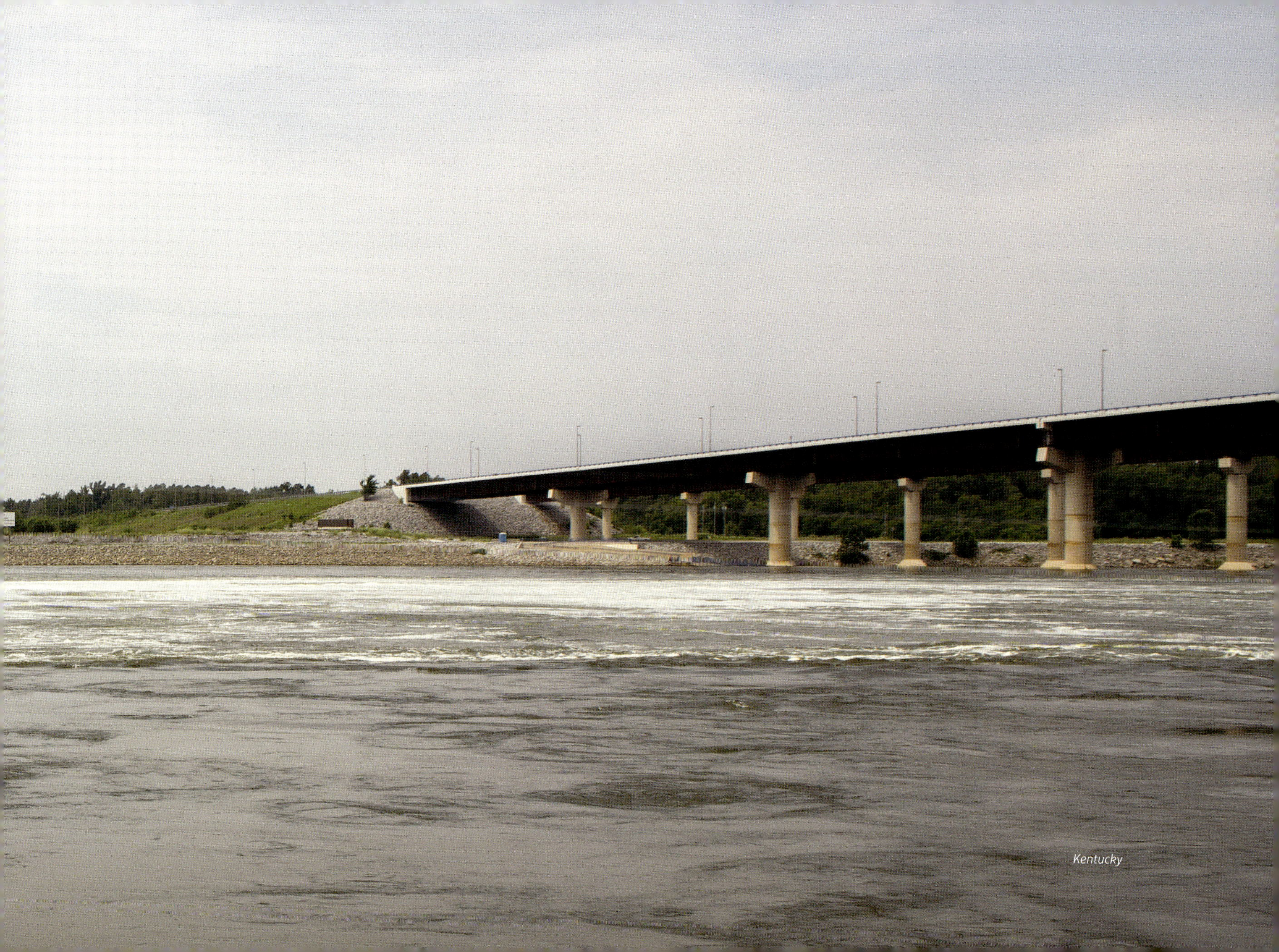
Kentucky

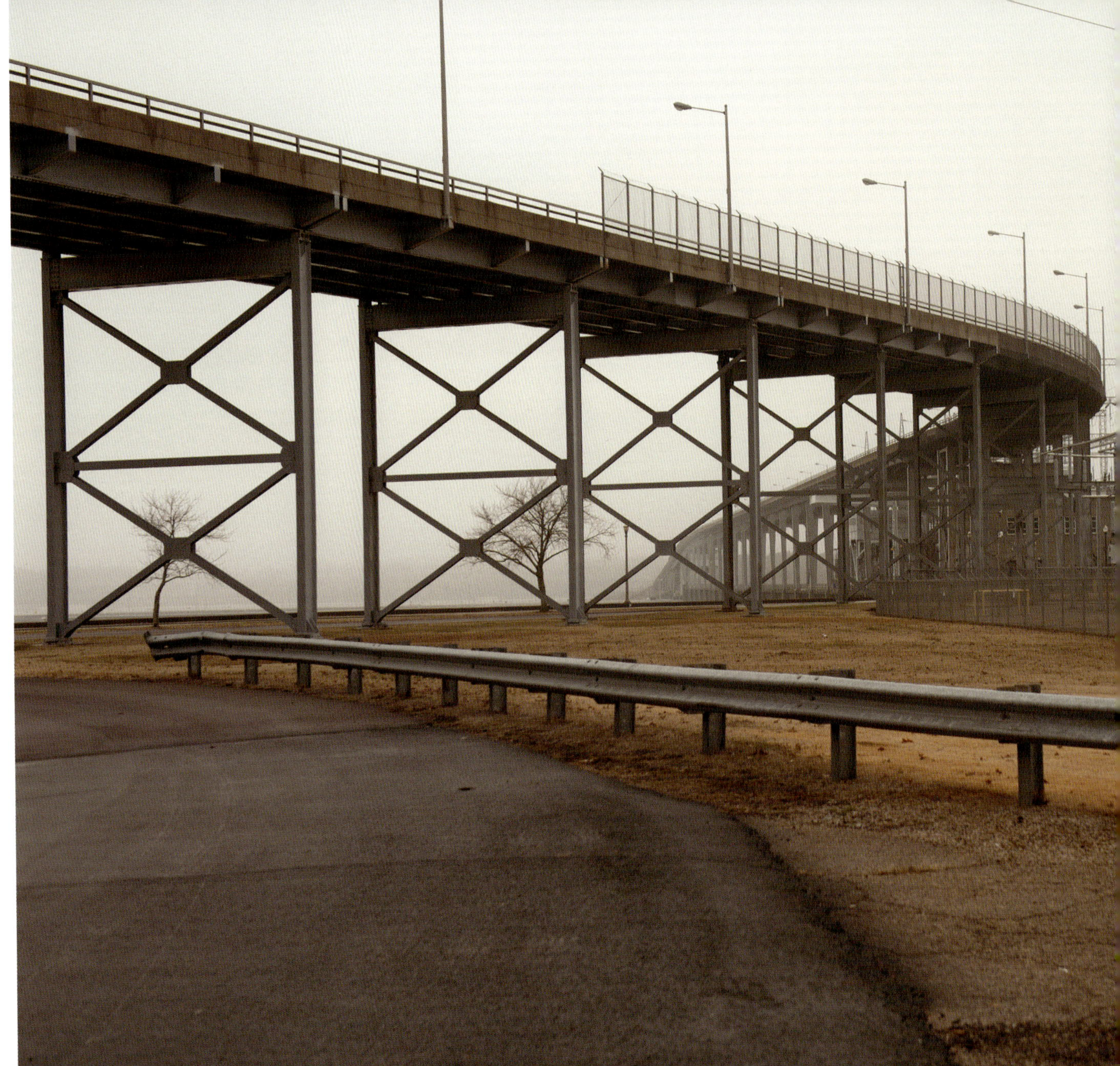

Pickwick Landing, Switchyard

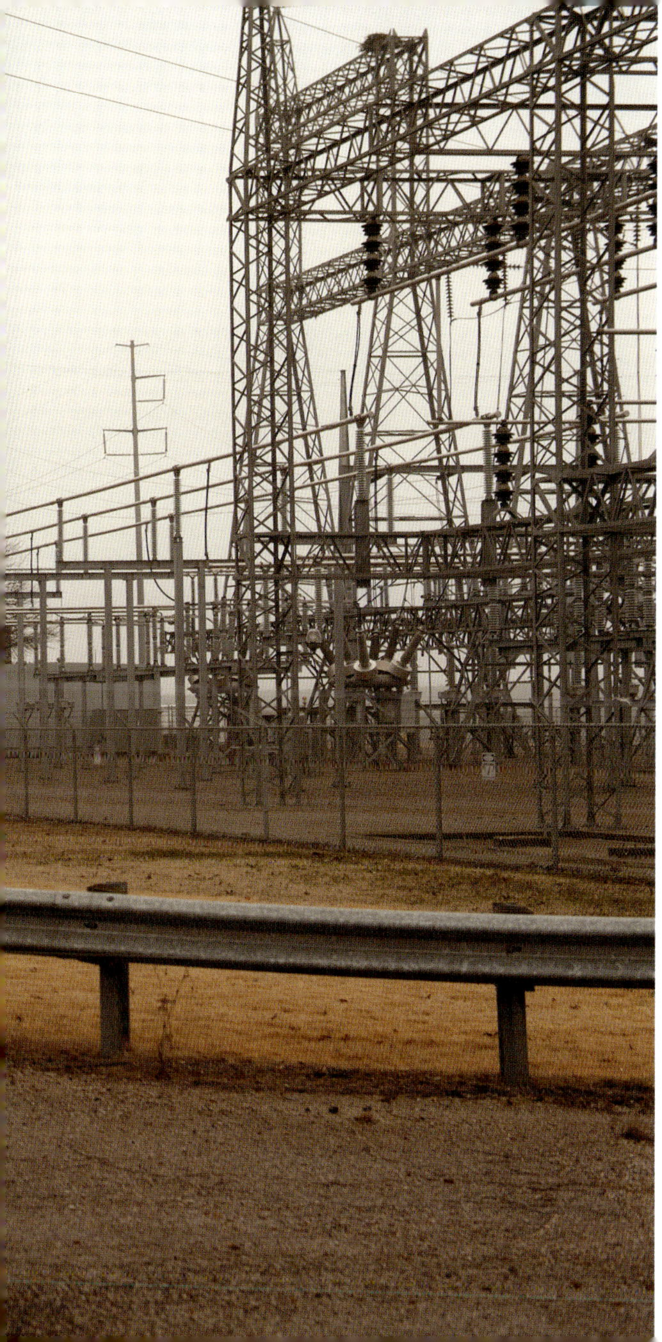

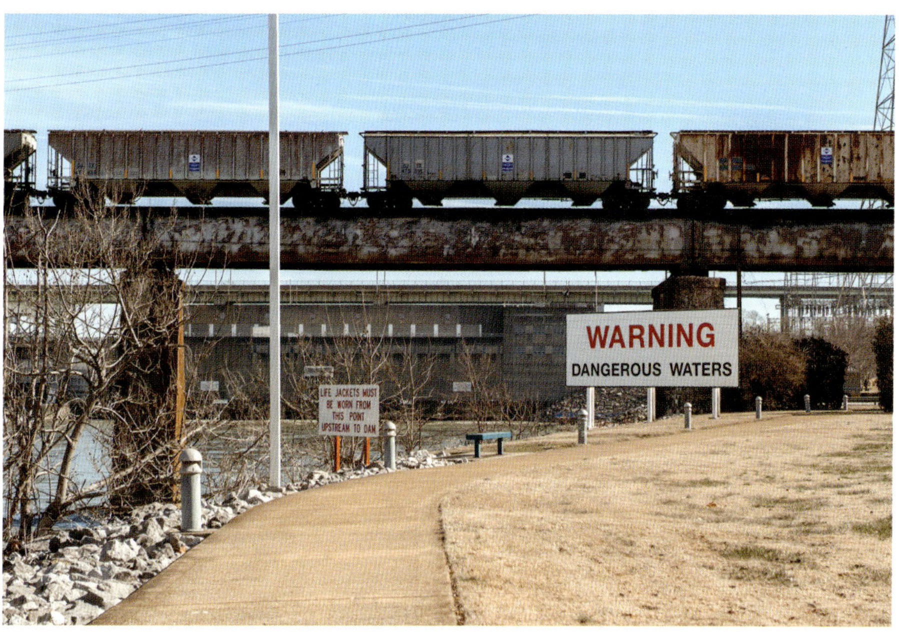

Chickamauga

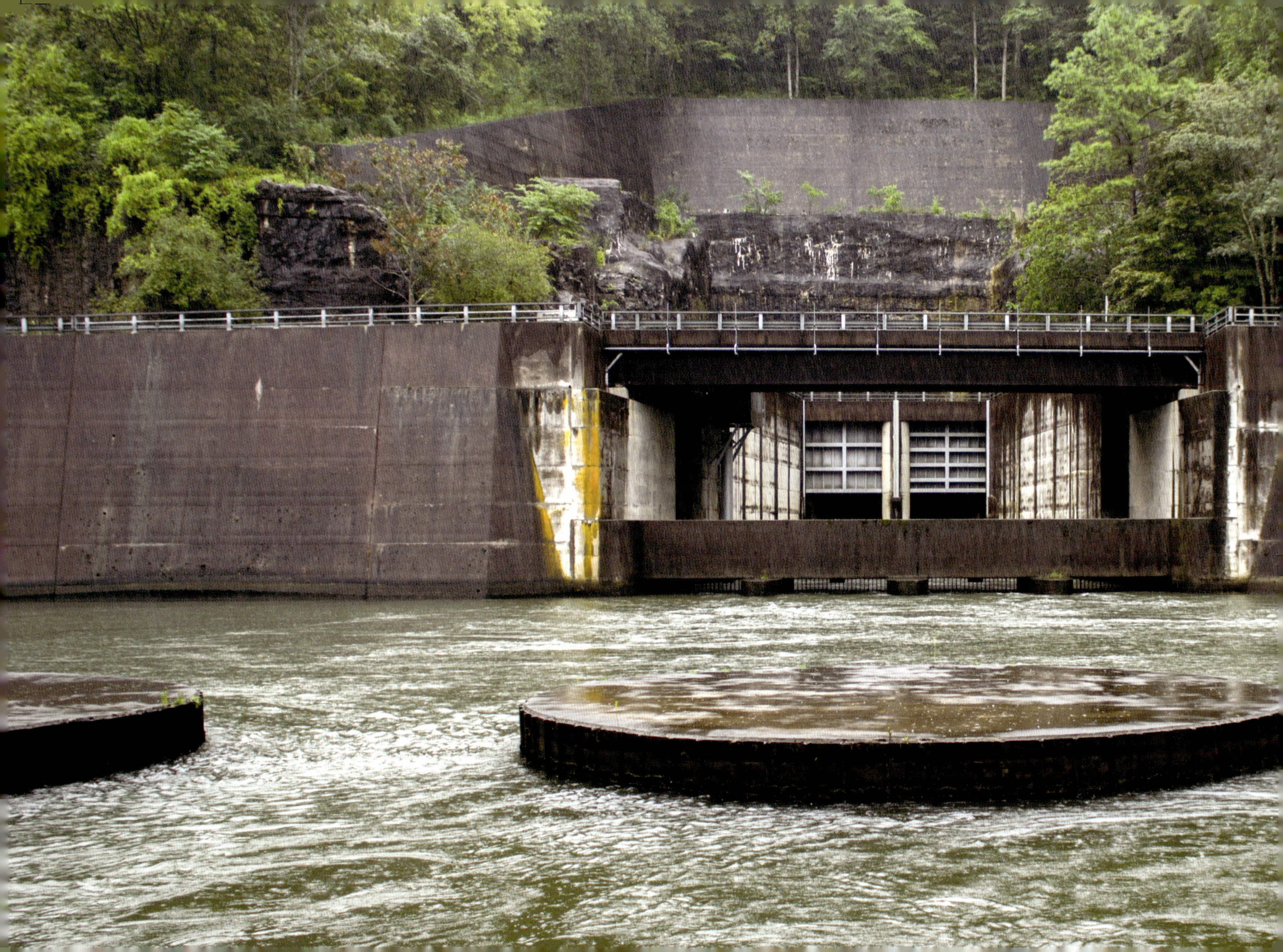

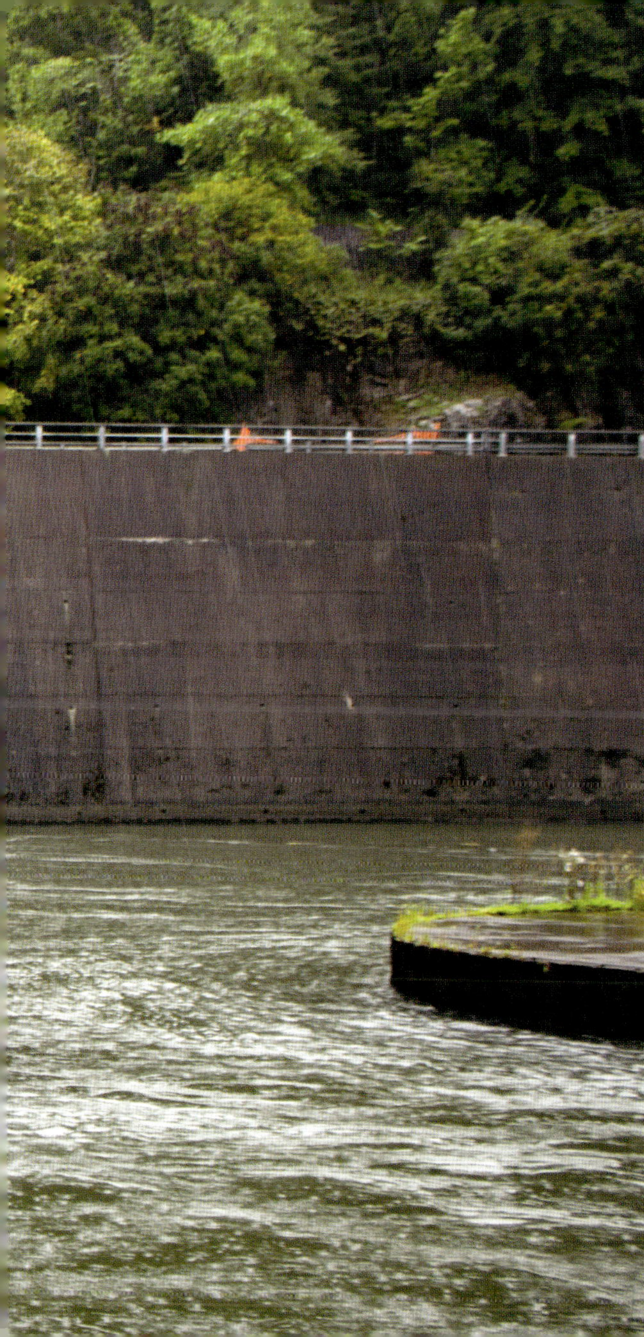

Raccoon Mountain, Discharge Tunnel

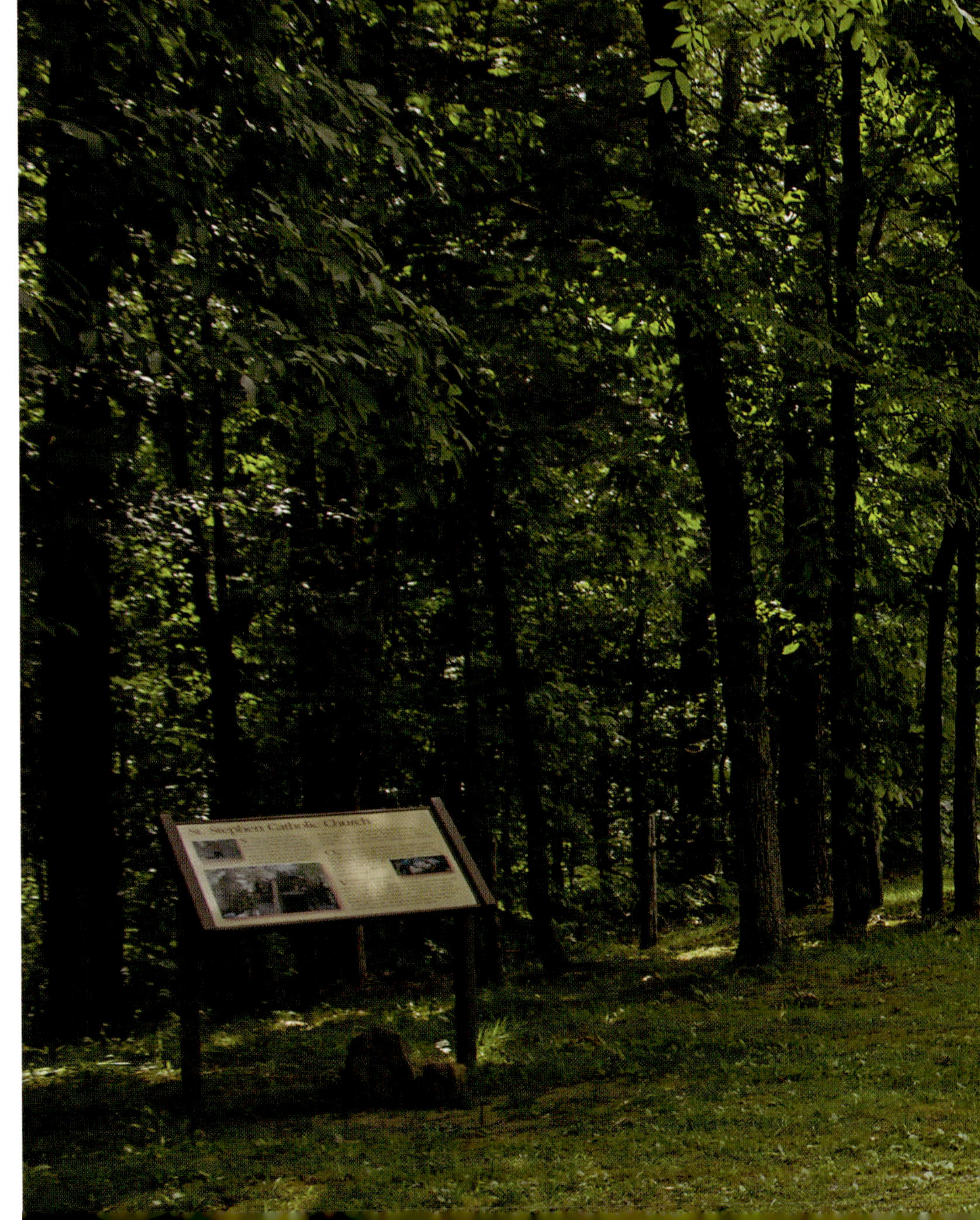

Land Between the Lakes, Saint Stephens Catholic Church

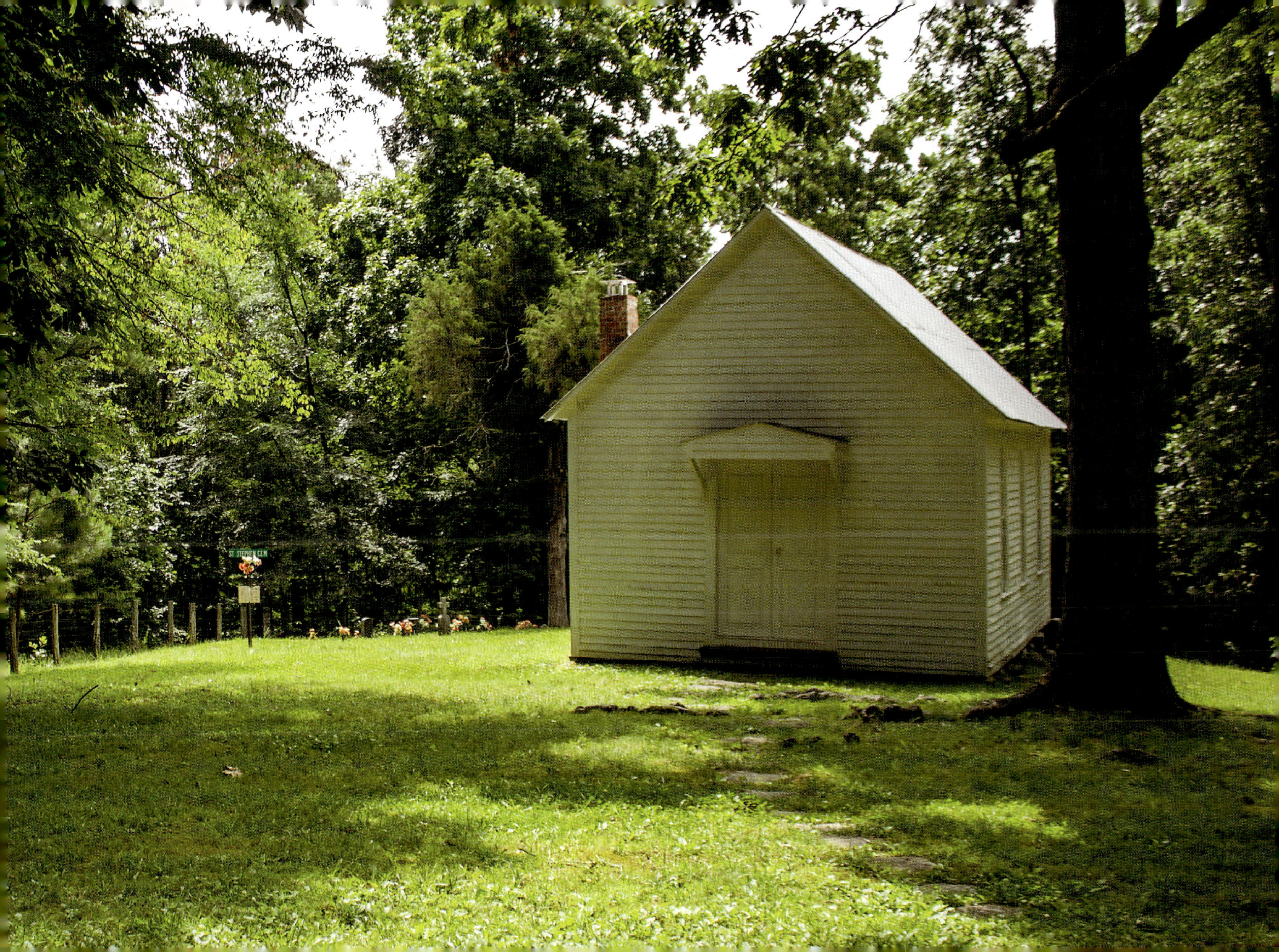

ST STEPHEN CEM

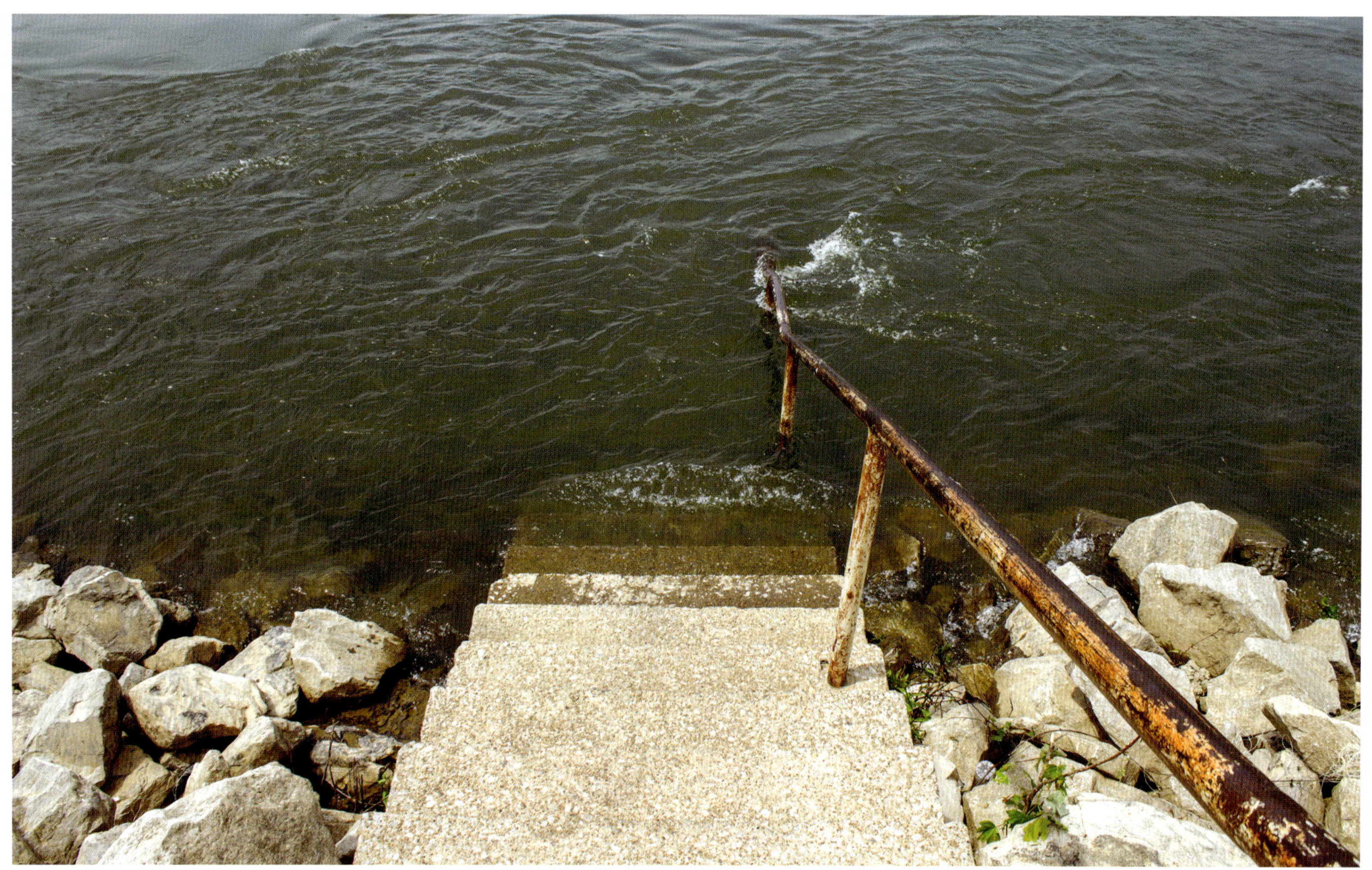

Kentucky, Downstream

I waded into *Dangerous Waters* out of curiosity, with a desire to explore these history-laden landscapes from a contemporary context and to document moments where industry and recreation coexist. More importantly, I was attracted to the narrative and purpose of TVA itself. I wanted to stand in the shadow of the dams and contemplate the continued commitment of the agency's social mission. I also wanted to look out over the reservoirs and testify on behalf of what was sacrificed. Every megawatt of hydroelectricity produced throughout TVA's service area is at the expense of dispossessed individuals. Yet, it is the continued mission of TVA—an agency

of service and economic development—that defines these dams as active enablers of its mission.

Some of the landscapes I explored and engaged with were filled with sounds of joy and community participation. Some were extremely well-manicured with updated recreational facilities and magnificent structures. And some I encountered held neglected architecture, inaccessible public space, and declared ecological modification. Many times, as many of my photographs show, these disparate visuals occurred at the same location.

The hydroelectric dams of the Tennessee Valley Authority do not operate in a vacuum. They coexist within a power portfolio containing other production methods such as coal, nuclear, natural gas, and renewable sources. While the dams constitute TVA's foundation and historical identity, they yielded their dominance to coal production half a century ago. It is important to keep in mind that the agency's social mission has advanced through the decades under this diversified umbrella of energy production. I purposely did not photograph these other programs because they require their own critical inquiry and thorough discussion.

While I firmly believe that the Tennessee Valley is better off thanks to TVA's initiatives, one must also respect the sacrifices made for that betterment. Exploring these locations with a camera—with the sole purpose of *seeing and observing*—allowed me to partially step into Valley residents' shoes while at the same time remaining a mere visitor. Because I stood there with a mind receptive to both sides of the visual, I empathize with those who lost their land to Tellico Dam. I understand how the dislocated and inconvenienced locals perceive TVA as a dishonest money-seeking entity out to aid private developers, but I can also appreciate how someone who is descended from a family that was displaced by Fontana Dam can be proud of a company that aided the war effort and improved living conditions in western North Carolina.

After years of conversations, research, and site visits, what I am left with is the agency's sincere attempt at public inclusion and communal well-being. As a whole, the physical and social costs of these dams are memorialized within TVA's identity and public correspondence. The narratives of displacement are not glossed over or withheld from conversation. Instead, they are edified within the visitor centers' informational exhibits and routinely referred to on TVA's website and social media accounts. In addition, the company's ongoing current-day concern for the public's experience must be mentioned. Many photographs in this book can attest to the fact that recreational facilities are being improved with newly built structures, and refurbished and reopened visitor centers are continuing to communicate TVA's electrical, ecological, and economic goals.

It is rare in our society to find a company as large and consequential as TVA that is not beholden to shareholder value. While significant mistakes have been made throughout its history, I believe that the same mission of regional development established in its charter continues to guide the organization, ensuring that such mistakes will not be repeated. However, as I stated in the introduction, *Dangerous Waters* is not about the past. While this body of work acknowledges the sacrifices and blunders made throughout TVA's long history, it is focused on the present and on how these facilities continue to operate in contemporary society. I believe that the very populace TVA exists to serve must hold the company accountable to its commitments. Those living within the TVA service area must maintain a watchful eye to ensure continued prosperity, opportunity, and an abundance of recreation are provided to the region.

Photography allowed me to not only record my experience but to document these spaces over a period of years. While I used the camera to interpret these locations from a variety of perspectives, the tonality was always my own. My journey throughout the service area began with wonder. The way the gray angles of the dams and their accompanying

structures jarred against emerald waters and forests was striking and beautiful. In the beginning, I was drawn to the arresting public presence of the architecture itself, but as time went on, I found myself more observant of the recreational spaces surrounding the dams. I began to question how people interacted with the landscape and photographed elements of TVA facilities that might normally go unnoticed: benches, picnic tables, windows, and restrooms.

The contemplation of historical and then contemporary dissonance gave way to acceptance of TVA's past and its shortcomings. I no longer approached these locations with suspicion but instead photographed them with understanding and admiration. Over the course of three years, I allowed myself to be captivated by TVA's dam architecture, saddened by its actions, skeptical of its motives, pleased with its accomplishments, moved by its mission, and supportive of its legacy. These reactions are not mutually exclusive, and to that many of the photographs within this book are examples.

My search for understanding and my landing in a place of acceptance can be symbolized by two photographs: *Kentucky, Downstream* (page 116) and *Land Between the Lakes, Saint Stephens Catholic Church* (page 114–115). These two images were made almost exactly three years apart. After weeks of travel in June of 2013, I found myself contemplating the history of TVA as I stood on those steps downstream of Kentucky Dam. I knew of the agency's progressive intention in its early years, of the determination and sacrifice illustrated by its wartime efforts, and of the dramatic social polarization of Tellico Dam. The weight of that history colored my view, and I made a photograph that symbolized my confusion and uncertainty.

For the next three years, I sought resolution—I did not want the project to end with the same level of unknowing with which it began. In the Land Between the Lakes National Recreation Area (LBL), I found what I was seeking. The LBL is a location that suffered three waves of eminent domain beginning with the construction of Kentucky Dam in the 1940s, the impounding of the Cumberland River by Barkley Dam in the 1950s, and ending with the total population removal of the peninsula in the 1960s. The area was not to be flooded, but was instead to be offered to the country as a multi-use recreational park. As was common practice, TVA either moved or completely dismantled all structures found on the purchased and condemned land, leaving the area's history solely up to social memory. Inexplicably though, the agency overlooked a single century-old Catholic church that had sat unused and empty since the 1940s. The structure was refurbished in 2001 by a group of former LBL residents as a symbol of who they were and what was once theirs.[1] This church stands as a place of gathering and remembrance for those who seek it out and is a testament to the power of community. I knew that my project was complete when I stood in the churchyard and created that final photograph.

Dangerous Waters is a visual record of my exploration and comprehension of TVA's consequential history and social presence. These images are not intended to speak in a collective voice. Instead, they are the product of my experience within these locations and offer my interpretive reaction to them. They are presented as components of a larger conversation about TVA, its history, and its continued mission of service. That conversation includes all of us, and I invite readers to visit these locations and come to their own conclusions. Hike the trails on Raccoon Mountain, raft down the Ocoee River, camp along the shores of Cherokee Lake, fish downstream of South Holston Dam, or stand on top of Norris Dam and see.

NOTE

1. Kimberly Hefling, "Church restored as memorial to those torn down by government," *Peninsula Clarion*, last modified January 19, 2001, http://peninsulaclarion.com.

NORTHEASTERN TRIBUTARIES

19   *Norris* (2013). Norris Dam was completed in 1936 and is located on the border of Anderson County and Campbell County in Tennessee. This photograph looks down Norris's spillway from atop the dam.

20   *Watauga, Reservoir* (2014). Watauga Dam was completed in 1948 and is located in Carter County, Tennessee. This photograph looks out over Watauga Lake and features one of Watauga's two intake towers.

21   *Melton Hill, Upstream* (2013). Melton Hill Dam was completed in 1963 and is located on the border of Roane County and Loudon County in Tennessee.

23   *Boone, Recreation Area* (2013). Boone Dam was completed in 1952 and is located on the border of Sullivan County and Washington County in Tennessee.

24   *Big Ridge State Park, Swimming Area* (2015). This photograph was made at Big Ridge State Park's public beach. This location within the park includes a number of swimming piers, a shallow wade pool, and a beach volleyball court.

25   *Boone, Visitor Center* (2013). This photograph shows the exterior of Boone Dam's visitor center. The building's glassed-in overlook was closed to the public.

27   *Norris, Downstream* (2013). This photograph was made along the Clinch River, not far from Norris Dam. The man remained fixated on the flowing current and did not acknowledge my existence as I made this photograph.

28   *South Holston, Downstream* (2014). These anglers are situated on a small weir downstream of South Holston Dam. This spot is popular for fishing and is adjacent to a public boat launch.

29   *Watauga, Visitor Center* (2014). Watauga Dam's overlook was closed to the public. The display panel

explains the history of the city of Butler, which was relocated to higher ground in 1948.

30 *Douglas, Visitor Center* (2014). This open-air overlook lies north of the dam at a large picnic area. Douglas Dam is located in Sevier County, Tennessee.

31 *Fort Patrick Henry, Visitor Center* (2013). Made when the facility was closed to the public, Fort Patrick Henry's visitor center overlooks the dam and its powerhouse. The facility was reopened in 2015.

32 *South Holston, Visitor Center* (2013). This photograph captures the back of South Holston Dam as seen through the windows of the closed visitor center.

33 *South Holston, Visitor Center (Reopened)* (2016). The wooden bench seen in the previous *South Holston, Visitor Center* photograph has been replaced by updated furniture and graphics. The building was refurbished and reopened in 2015.

34 *Cherokee, Recreation Area* (2014). This photograph was made near Cherokee Dam's public beach. Cherokee Dam is located in Grainger and Jefferson Counties in Tennessee. It was completed in 1941.

35 *Big Ridge State Park* (2015). Big Ridge State Park was created by TVA, the Civilian Conservation Corps, and the National Park Service as a demonstration recreational area. It was a component of the larger Norris Project and opened in 1934.

36 *Norris, Marina* (2013). Norris Dam Marina is located near this pictured quarry.

37 *Wilbur* (2013). Wilbur Dam is located approximately three miles upstream from Watauga Dam in Carter County, Tennessee. It was completed in 1912 and acquired by TVA in 1945.

38 *South Holston, Infuser Weir* (2014). The pictured weir is a popular location for picnickers and anglers. It is located downstream from South Holston Dam at Osceola Island.

41 *Douglas, Recreation Area* (2014). This public beach is located on Douglas Lake near one of the dam's camping facilities.

42 *South Holston* (2014). Located in Sullivan County, Tennessee, South Holston Dam was completed in 1950. This photograph was made at Stophel Cemetery located downstream from the dam.

SOUTHEASTERN TRIBUTARIES

51 *Ocoee #2* (2014). Ocoee Dam No. 2 was completed in 1913 and acquired by TVA in 1939. It is located in Polk County, Tennessee.

52 *Fontana* (2014). Fontana Dam is the tallest dam in the eastern United States and was completed in 1945. It is located in Swain County and Graham County in North Carolina. The dam's reservoir borders Great Smoky Mountains National Park.

53 *Apalachia, Reservation* (2015). Apalachia Dam is located in Cherokee County, North Carolina. It was completed in 1943.

54   *Blue Ridge* (2014). Blue Ridge Dam is located in Fannin County, Georgia, and was completed in 1930. The top of the dam and its intake tower are pictured in the photograph.

55   *Chatuge, Reservoir* (2014). A family swims along the embankment near Chatuge Dam.

56   *Chatuge, Infuser Weir* (2014). The pictured weir is located downstream of Chatuge Dam.

57   *Fontana, Reservoir Overlook* (2015). This pavilion is located near the dam's visitor center and was in the process of being decorated for a wedding ceremony.

58   *Hiwassee, Reservoir Overlook* (2014). This facility is located near the top of Hiwassee Dam and looks out over its reservoir. It includes basic facilities and a few educational display panels.

58   *Blue Ridge, Picnic Area* (2014). This small picnic area overlooks Blue Ridge Dam and its powerhouse. Pictured are two historical markers and an educational display discussing the refurbishment of the dam.

59   *Hiwassee* (2014). Hiwassee Dam is located in Cherokee County, North Carolina, and was completed in 1940. The pictured gantry crane is located at the top of the dam.

60   *Chatuge* (2014). Chatuge Dam is located in Clay County, North Carolina, and was completed in 1942.

61   *Apalachia* (2015). The pictured penstock carries water to the dam's powerhouse in Polk County, Tennessee.

62   *Nottely* (2014). Nottely Dam is located in Union County, Georgia, and was completed in 1942. The dam's spillway is pictured.

63   *Ocoee #1* (2015). Ocoee Dam No. 1 is located in Polk County, Tennessee, and was completed in 1911. This photograph was made directly downstream of the dam at Sugarloaf Mountain State Park.

64   *Fontana, Visitor Center* (2013). This photograph was made from inside Fontana Dam's visitor center.

66   *Tellico, Reservoir* (2013). This photograph was made near one of Tellico Dam's public beaches.

66   *Fontana, Lakeview Drive* (2015). The pictured pylons are found within Great Smoky Mountains National Park and mark the end of Lakeview Drive.

67   *Ocoee #3, Powerhouse* (2016). Water from Ocoee Dam No. 3 is carried two and a half miles by this penstock. Ocoee Dam No. 3 is located in Polk County, Tennessee, and was completed in 1943. The dam's powerhouse was behind me when I made this photograph.

CENTRAL AND WESTERN RESERVATIONS

73   *Great Falls* (2015). Great Falls Dam is located in Warren County and White County, Tennessee. It was completed in 1916.

74   *Little Bear Creek, Reservoir* (2014). Little Bear Creek Dam is located in Franklin County, Alabama, and was completed in 1975. The dam's intake tower is pictured.

75   *Normandy* (2016). Normandy Dam is located on the border of Bedford County and Coffee County in Tennessee. It was completed in 1976.

76 *Tims Ford* (2014). Tims Ford Dam is located in Franklin County, Tennessee. It was completed in 1975.

77 *Cedar Creek* (2015). Cedar Creek Dam is located in Franklin County, Alabama. It was completed in 1979. The dam's outlet works are pictured.

78 *Bear Creek, Recreation Area* (2016). This photograph was made upstream from Bear Creek Dam at a seldom used and overgrown public beach. The dam and its intake tower can be seen in the photograph.

79 *Upper Bear Creek, Reservoir* (2016). Upper Bear Creek is located in Marion County, Alabama, and was completed in 1978. The dam's intake tower is pictured.

79 *Bear Creek* (2014). Bear Creek Dam is located in Franklin County, Alabama. It was completed in 1969.

81 *Little Bear Creek, Fishing Pier* (2016). This large fishing pier is located on the dam's reservoir.

83 *Dogwood* (2014). Dogwood dam is located on the Beech River in Covington County, Alabama. It was completed in 1965 and does not contain any hydroelectric components.

92 *Fort Loudoun, Downstream*, (2013). Fort Loudon Dam is located in Loudon County, Tennessee. It was completed in 1943. This photograph looks toward the dam from a downstream recreational facility.

93 *Pickwick Landing, Downstream* (2014). Pickwick Landing Dam is located in Hardin County, Tennessee. It was completed in 1938.

94 *Wheeler* (2014). Wheeler dam is located on the border of Lauderdale and Lawrence Counties in Alabama. It was completed in 1936.

95 *Raccoon Mountain* (2015). This structure surrounds the intake/discharge tunnel at the foot of Raccoon Mountain Pumped Storage Plant. The facility is located on the Tennessee River in Marion County, Tennessee, and was completed in 1978.

96 *Nickajack* (2014). Nickajack Dam was completed in 1967 and was built to replace nearby Hales Bar Dam. It is located in Marion County, Tennessee.

97 *Wilson, Downstream* (2015). Wilson Dam is located on the border of Lawrence County and Colbert County in Alabama. It was completed in 1924 and was transferred to TVA's ownership in 1933. The photograph shows a group of anglers directly downstream from Wilson's powerhouse. A "Dangerous Waters" billboard rises above them.

98 *Kentucky, Fishing Pier* (2013). This fishing pier is located downstream from Kentucky Dam's powerhouse.

99 *Fort Loudoun, Upstream* (2016). This permanently closed facility is located upstream from Fort Loudoun's powerhouse and in close proximity to Lenoir City Park. It still contains a number of picnic tables, but the bathroom facilities and public beach are overgrown

and unusable. The sign points visitors to the nearby beach facility at Tellico Dam.

100 *Fort Loudoun, River Access* (2013). This sign was photographed near Fort Loudoun Dam. It warns against contaminated fish that are found in this body of water

100 *Watts Bar, Visitor Center* (2013). This photograph was made on the pedestrian bridge that crosses over TN-68. The bridge connects the dam's visitor center to its parking lot.

101 *Wilson, Visitor Center* (2014). This security gate is located near recently installed display panels that serve as Wilson Dam's official "visitor center."

102 *Guntersville* (2016). Guntersville Dam is located in Marshall County, Alabama. It was completed in 1939.

104 *Chickamauga, Downstream* (2014). This bench offers a place of respite and observation downstream from Chickamauga Dam's powerhouse.

105 *Wilson, Recreation Area* (2015). This playground is located in Rockpile Recreation Area.

106 *Watts Bar, Recreation Area* (2013). This photograph was made at a public beach near Watts Bar Dam. Watts Bar Dam is located on the border of Meigs and Rhea Counties in Tennessee and was completed in 1942. The cooling towers of Watts Bar Nuclear Plant can be seen in the background.

107 *Hales Bar, Powerhouse* (2016). Hales Bar Dam ceased operation in 1967 and was dismantled following the completion of nearby Nickajack Dam. The former powerhouse now serves as dry boat storage for Hales Bar Marina.

108 *Kentucky* (2013). Kentucky Dam is located in Livingston and Marshall Counties in Kentucky. It was completed in 1944.

110 *Pickwick Landing, Switchyard* (2014). The dam's switchyard is situated next to the dam's powerhouse and parking lot.

111 *Chickamauga* (2014). Chickamauga Dam is located in Hamilton County, Tennessee. It was completed in 1940.

112 *Raccoon Mountain, Discharge Tunnel* (2015). Raccoon Mountain Pumped Storage Plant contains a mountaintop reservoir that is drained during peak electricity demand. The pictured discharge tunnel also operates as the intake location when the reservoir is refilled.

115 *Land Between the Lakes, Saint Stephens Catholic Church* (2016). Land Between the Lakes National Recreation Area was created in 1963. The peninsula of land is created by Kentucky Lake to the west and Lake Barkley to the east. Prior to the construction of Kentucky Dam and Barkley Dam, the region was known as the Land Between the Rivers.

116 *Kentucky, Downstream* (2013). This stairway leads to a concrete fishing pier. Kentucky Dam was spilling water when I made this photograph and the pier was currently inaccessible due to the Tennessee River's high water.

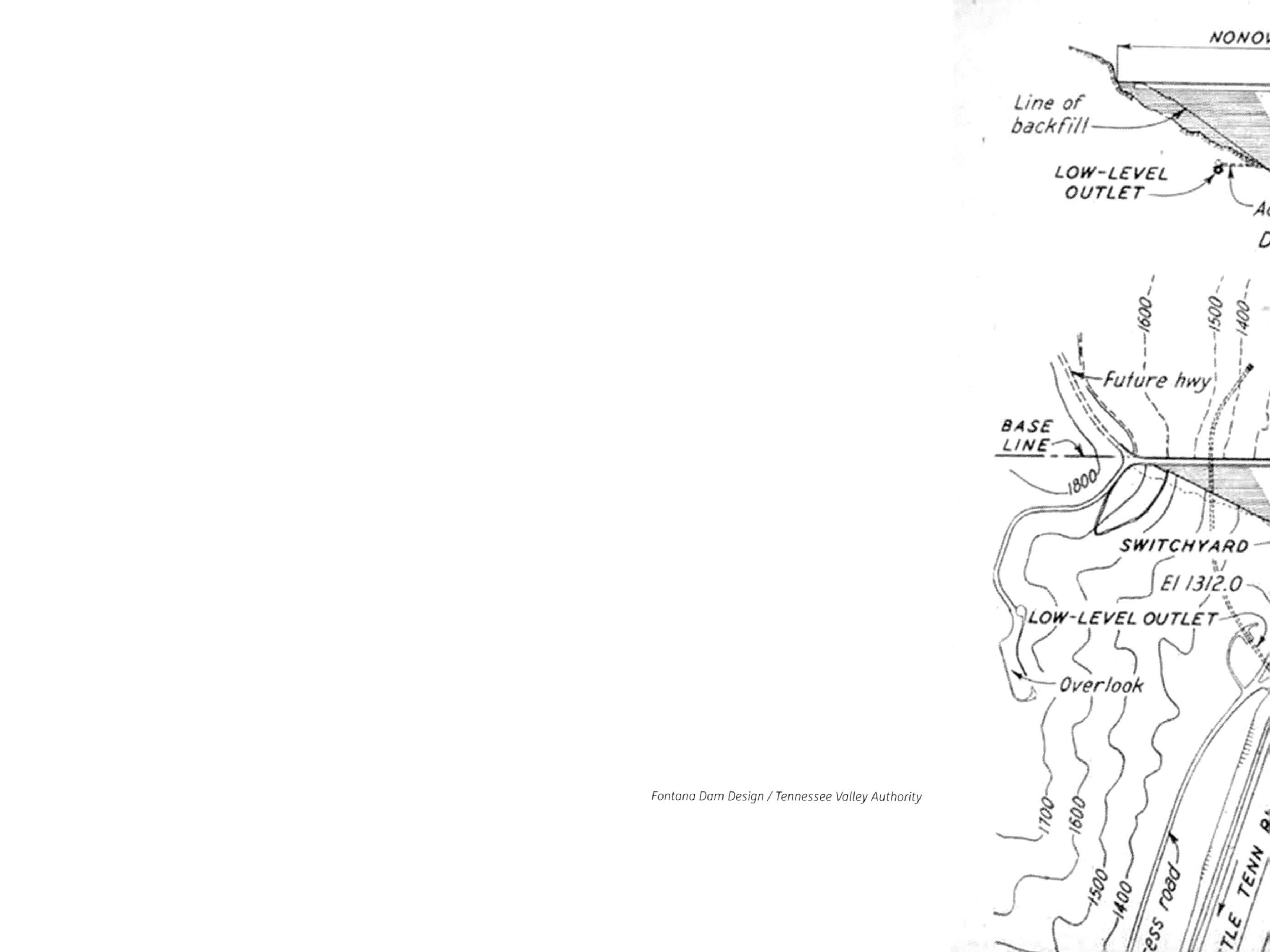

Fontana Dam Design / Tennessee Valley Authority

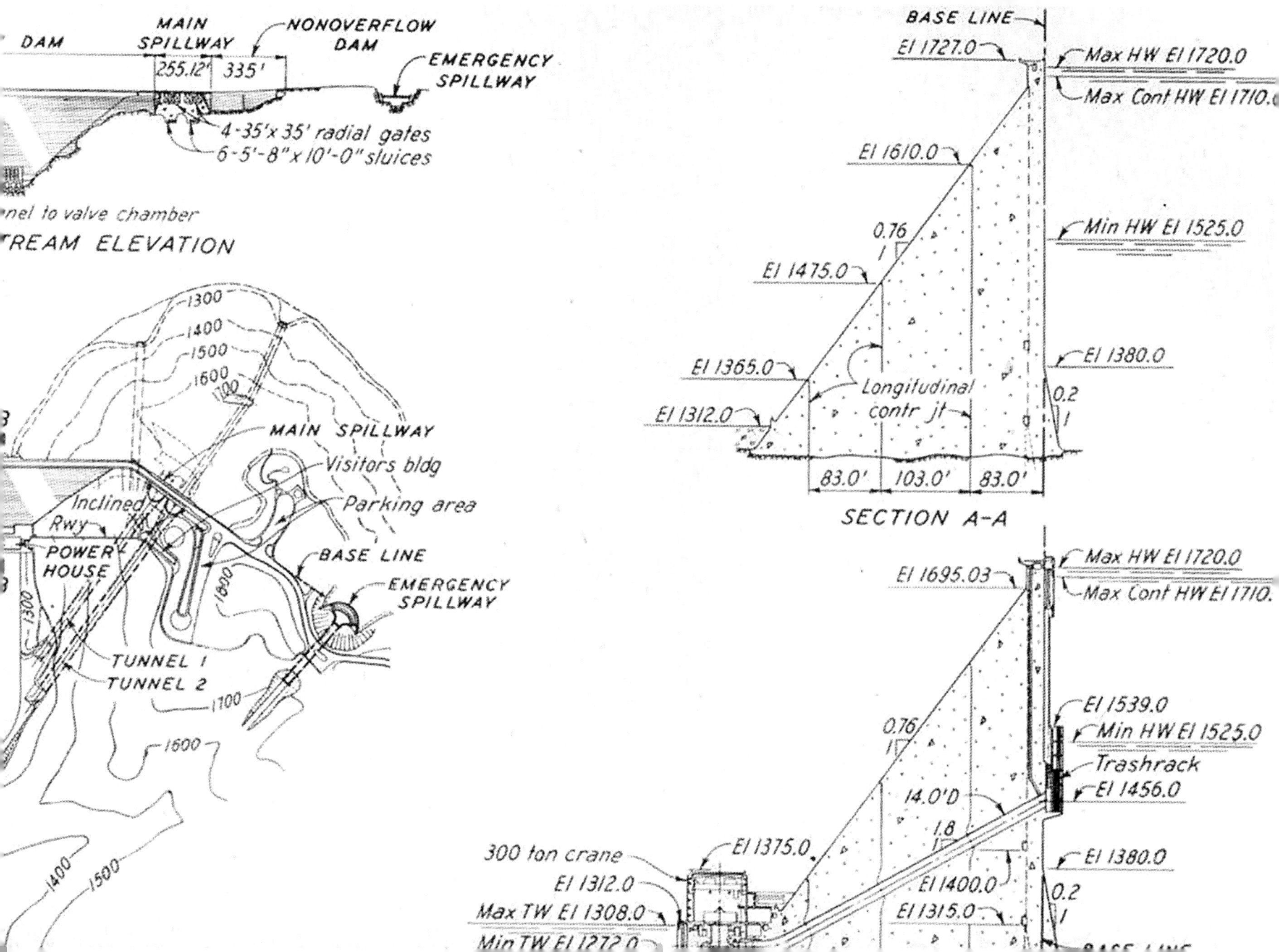

DAM
MAIN SPILLWAY
255.12'   335'
NONOVERFLOW DAM
EMERGENCY SPILLWAY
4-35'x 35' radial gates
6-5'-8"x 10'-0" sluices
nel to valve chamber
TREAM ELEVATION
1300
1400
1500
1600
MAIN SPILLWAY
Visitors bldg
Parking area
Inclined Rwy
POWER HOUSE
BASE LINE
EMERGENCY SPILLWAY
1800
1300
TUNNEL 1
TUNNEL 2
1100
1600
1400
1500
BASE LINE
El 1727.0
Max HW El 1720.0
Max Cont HW El 1710.0
El 1610.0
Min HW El 1525.0
El 1475.0
0.76
1
El 1365.0
El 1312.0
Longitudinal contr jt
El 1380.0
0.2
1
83.0'   103.0'   83.0'
SECTION A-A
El 1695.03
Max HW El 1720.0
Max Cont HW El 1710.
0.76
1
El 1539.0
Min HW El 1525.0
Trashrack
El 1456.0
14.0'D
1.8
300 ton crane
El 1375.0
El 1312.0
Max TW El 1308.0
Min TW El 1272.0
El 1400.0
El 1315.0
El 1380.0
0.2
1
BASE LINE

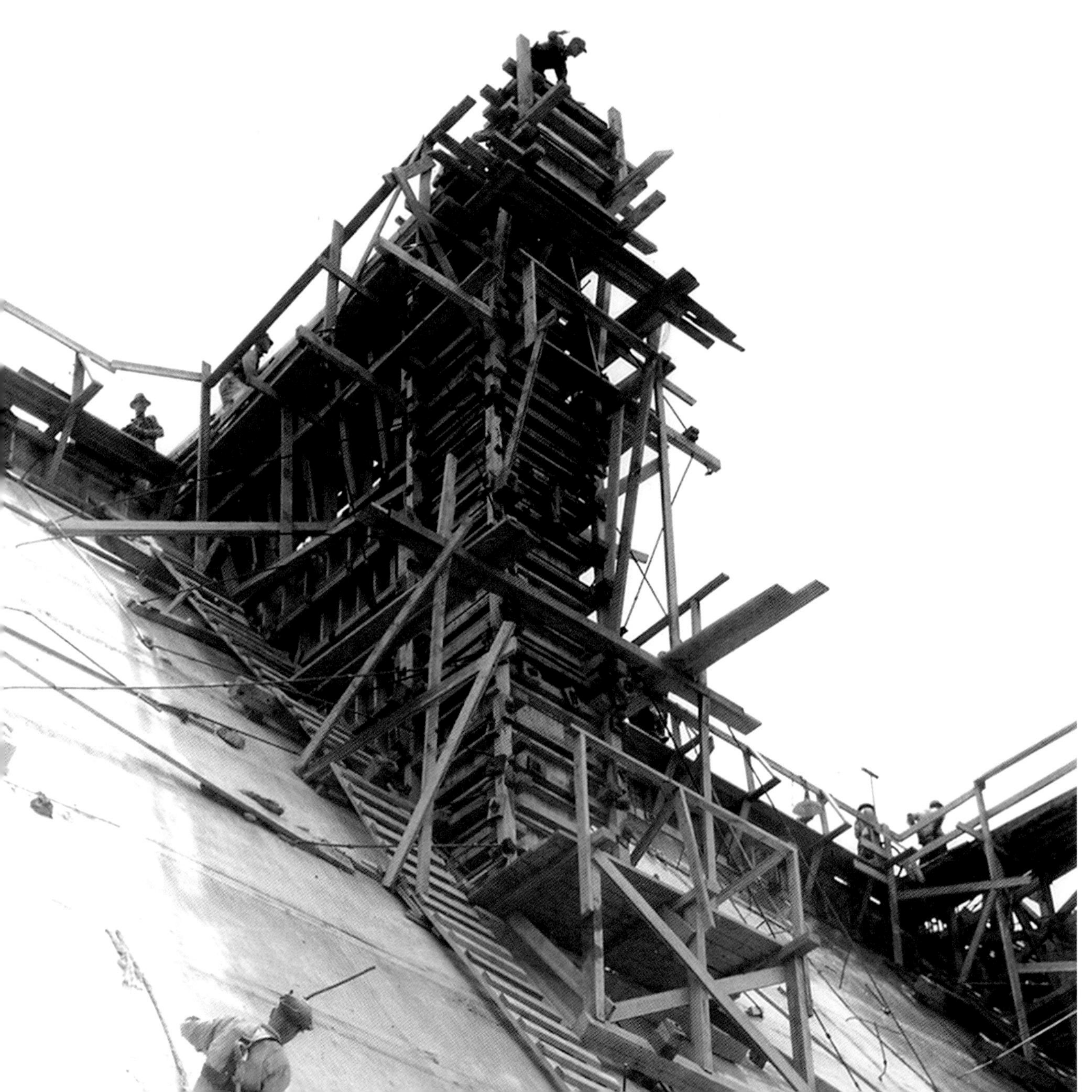

Wheeler Dam (1935) Construction Worker

/ Tennessee Valley Authority